Patient To Provider

A Patient's Guide To Transplantation

by

Jeffrey W. Young, Jr. MSN FNP APN-BC.

authorHOUSE™

1663 LIBERTY DRIVE, SUITE 200
BLOOMINGTON, INDIANA 47403
(800) 839-8640
WWW.AUTHORHOUSE.COM

First published by AuthorHouse 03/02/05

ISBN: 1-4208-3795-8 (sc)

Printed in the United States of America
Bloomington, Indiana

This book is printed on acid-free paper.

*Dedication: This book is dedicated
to all organ donors, living and deceased,
and their family for giving the ultimate gift...
The Gift of Life!*

In honor of my Mother, Barbara, who gave me the Gift of Life, twice! Thanks Mom, I love you.

To my loving wife Holly, only I know the sacrifices made. I could not be what I am without you. You are truly the secret to my success... yours always, with all my love....Jeff

TABLE OF CONTENTS

———— ⚜ ————

Introduction ix

Preface *"Never give up!"* xi

Introduction to Transplantation 1

Dealing with your Diagnosis 2

From Patient to Provider. 10

What is Kidney Failure? 18

What is Heart Failure? 30

What is Liver Failure? 40

What is Pancreas Failure? 53

What is Lung (Pulmonary) Failure? 63

Life after Transplantation 73

Stories of Success 82

Pediatric Transplant Success Stories 109

Frequently Asked Questions 114

Pediatric FAQ 123

Common Myths Of Organ Donation 129

Bibliography 132

INTRODUCTION

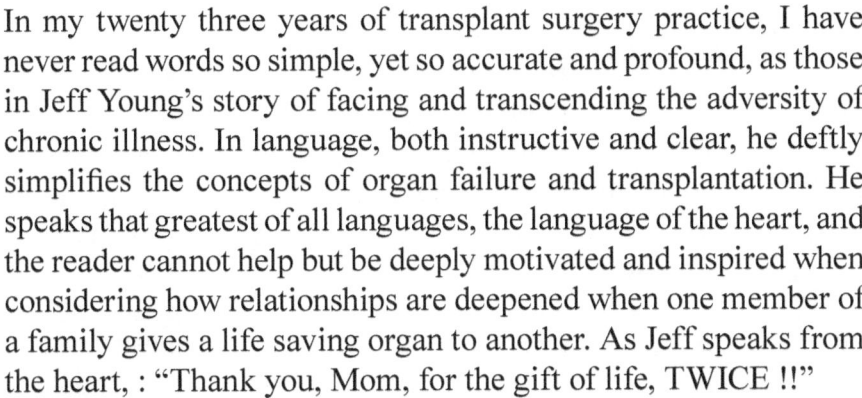

In my twenty three years of transplant surgery practice, I have never read words so simple, yet so accurate and profound, as those in Jeff Young's story of facing and transcending the adversity of chronic illness. In language, both instructive and clear, he deftly simplifies the concepts of organ failure and transplantation. He speaks that greatest of all languages, the language of the heart, and the reader cannot help but be deeply motivated and inspired when considering how relationships are deepened when one member of a family gives a life saving organ to another. As Jeff speaks from the heart, : "Thank you, Mom, for the gift of life, TWICE !!"

The personal portrait of an eighteen year old, denying his symptomatology until the very end of his renal function, this self-searching work looks deep into the radical changes in self-concept brought about by the realization that he is no longer a "healthy" man, but one who will always require medicines and doctor's visits. While it is true that in my practice of organ transplantation I can provide a state of health quite close to normal, it is not the same, and there are deep adjustments to be made on the part of the patient. The book's dual perspectives of Jeff's personal experiences, both as a patient and a health care provider, along with the skillful clinical descriptions of the many types of organ failure, provide the reader with a sound foundation for dealing with that moment in life when a patient is told he/she is no longer NORMAL.

The didactic section instructs on how the body functions when organs are damaged, explains how transplants are performed, outlines the warning signs and symptoms of complication, and presents a concrete plan for preventing and addressing such complications. Here Jeff has masterfully found the balance between clarity and depth.

The final section in which public statements from transplant recipients give a name and face to the many patients helped by live and deceased donor transplantation, is truly inspirational.

It is my hope that in treating very ill patients for so many years God has used me as an intermediary between Him and the sick, granting me in his grace the loan of His wisdom, compassion and technical ability for the benefit of the suffering. The "never give up" stories Jeff tells have been my way of lifting my patients when they are low.

Above anything, this book is a testament to how God uses us, His care givers, to minister to the sick and to reveal His unfailing love for them even whey they face their greatest trial. This book is a gift to us all.

Santiago R. Vera FACS
Director Liver Transplantation
University of Tennessee, Memphis
(Jeff's doctor and FRIEND!)

PREFACE

"Never give up!"

*"Life is 10% of what happens to you and 90%
of how you react to it." – Charles Swindoll*

On October 29, 1941, U.K. Prime Minister Winston Churchill visited Harrow School to hear the traditional songs he had sung there as a youth, as well as to speak to the students. This became one of his most quoted speeches. The myth is that Churchill stood before the students and said, "Never, ever, ever, ever, ever, ever, ever, give up. Never give up. Never give up. Never give up." Then he sat down.

In reality, he made a complete speech that included words similar to what are often quoted. "Never give in. Never give in. Never, never, never, never--in nothing, great or small, large or petty--never give in…"

Whether he spoke only the words "never give up" and sat down or he gave a whole speech and that included the part about "never give in" the debate doesn't matter the content and truth of what he said is powerful. This book is about people who were dealt a deck of cards in life to play with and the played them…well. But the thread that is the same throughout is the truth that Churchill spoke on that fall day in Great Britain…never give up!

Bad things do happen to good people. This is a fact of life. But you must never give up! There are always trials and tests, but it is up to <u>us</u> whether or not they become <u>obstacles</u> or <u>opportunities</u>.

In ancient times, a king had a boulder placed on a roadway. Then he hid himself and watched to see if anyone would remove the huge rock from the road. Some of the king's wealthiest merchants and courtiers came by and simply walked around it. Many loudly blamed the king for not keeping the roads clear, but none did anything about getting the stone out of the way. Then a peasant came along carrying a load of vegetables. Upon approaching the boulder, the peasant laid down his burden and tried to move the stone to the side of the road. After much pushing and straining, he finally succeeded. After the peasant picked up his load of vegetables, he noticed a purse laying in the road where the boulder had been. The purse contained many gold coins and a note from the king indicating that the gold was for the person who removed the boulder from the roadway. The peasant learned what many of us never understand. Every obstacle presents an opportunity to improve our condition.

Never give up!

Benjamin Disraeli, a novelist, a brilliant debater and England's first and only Jewish prime minister, Disraeli (Earl of Beaconfield) is best remembered for bringing India and the Suez Canal under control of the crown stated, "There is no education like adversity."

Never give up!

I read a book by Jim Rohn, philosopher, genius, motivational counselor, and business executive --- I will never forget what he said:

> Success is neither magical nor mysterious.
> Success is the natural consequences of consistently applying the basic fundamentals.

Never give up!

Everything that has gotten you to this point in your life is all you need to make you successful in adversity. We have all had to overcome little obstacles in our life, and what got us through those minor altercations are the same principles that it takes to overcome a major setback in life.

Although success is nothing more than a few simple disciplines practiced everyday, failure is simply a few errors in judgment repeated everyday.

It is the accumulated weight of our disciplines and our judgments that lead us to success or failure!

<u>Never give up</u>!

Success is not so much what we have, as it is, what we are.

<u>Never give up</u>!

Perseverance and hard work will accomplish anything!

<u>Never give up</u>!

No one's life better exemplifies this precept than the life of Abraham Lincoln.

In 1831 -- failed at business
   1832 -- defeated for Legislature
   1833 -- again failed in business
   1834 -- elected to Legislature
   1835 -- childhood sweetheart died
   1836 -- had a nervous breakdown
   1838 -- defeated for speaker of the house
   1840 -- defeated for elector
   1843 -- defeated for Congress
   1846 -- elected to Congress
   1848 -- defeated again for Congress
   1855 -- defeated for Senate
   1856 -- defeated for Vice President
   1858 -- defeated again for Senate
   1860 -- elected President of the United States

Never give up!

Throughout his life, Abraham Lincoln suffered much more defeat than victory, but in the end-- it is his victories that are remembered.

I'm sure there were many times that he felt like giving up -- yet he persevered, and today he is remembered as one of the greatest Presidents that ever lived.

Never give up!

There is one book in the Bible <u>dedicated</u> to this concept--

The book of Job. Talk about a guy who faced adversity, he lost everything. He was very wealthy, had a family, and good health. And he lost everything including his health. "Though he slay me, yet will I serve Him" was his attitude.

Job never gave up! And in the final analysis became more successful than he was before his adversity.

Success is neither magical nor mysterious. Success is the <u>natural consequences</u> of consistently applying <u>basic fundamentals</u>!

Never give up!

As you read this book I hope you will learn something that will help you get through this time in your life. Every person mentioned in this book has been where you are, felt the feelings you are feeling, experienced the uncertainty of tomorrow, yet are mentioned in this book because they never gave up.

No matter how tough it gets, or how uncertain tomorrow seems... never, never, never give up!!

---Jeffrey W. Young, Jr., MSN, FNP, APN-BC

INTRODUCTION
TO TRANSPLANTATION

What is transplantation?

Transplantation takes place when an organ from one person is surgically removed, and placed into another person. It is a treatment in situations where a person's organ has failed because of illness or injury. Replacing the organ may be the only treatment choice for the patient or the best among several options.

What organs can be transplanted?

Solid organs that are transplanted include:

- heart
- lung
- liver
- kidney
- pancreas
- intestine

In some cases, two organs are transplanted at once. Examples of this are:

- heart-lung
- kidney-pancreas

Tissues that can be transplanted include:

- cornea
- bone
- cartilage
- skin
- heart valves
- saphenous vein

DEALING WITH YOUR DIAGNOSIS

──────── ❖ ────────

Living with a long-lasting health condition (also called a chronic illness) presents a person with new challenges. Learning how to meet those challenges is a process - it doesn't happen right away. But understanding more about your condition, and doing your part to manage it, can help you take health challenges in stride. Many people find that taking an active part in the care of a chronic health condition can help them feel stronger and better equipped to deal with lots of life's trials and tribulations.

If you're living with a chronic illness, you may feel affected not just physically, but also emotionally, socially, and sometimes even financially. The way a person might be affected by a chronic illness depends on the particular illness and how it affects the body, how severe it is, and the kinds of treatments that might be involved. It takes time to adjust to and accept the realities of a long-term illness, but teens who are willing to learn, seek support from others, and participate actively in the care of their bodies usually get through the coping process.

Coping With a Chronic Illness

Coming to terms with a chronic illness is a trying process that takes a lot of time, energy and effort. It's a journey that can take you through a range of emotions — including grief, anger, relief (at finally knowing what is wrong), fear and hope — and potentially force you to take stock of your life, the people in it and the world around you. These feelings are the start of the coping process. Everyone's reaction is different, but they're all completely normal.

Elizabeth Kubler-Ross identified five stages of grief that you may experience when your first given your diagnosis. The stages Kubler-Ross identified are: Denial (this isn't *happening* to me!),

Anger (why is this happening to *me*?), Bargaining (I promise I'll be a better person *if...*), Depression (I don't *care* anymore), and Acceptance (*I'm ready* for whatever comes).

Emotions may not be easy to identify. For example, sleeping or crying a lot or grouchiness may be signs of sadness or depression. It's also very common for people with chronic illnesses to feel stress as they balance the realities of dealing with a health condition and coping with schoolwork, social events, and other aspects of everyday life.

Many people living with chronic illnesses find that it helps to line up sources of support to deal with the stress and emotions. Some people choose to talk to a therapist or join a support group specifically for people with their condition. It's also important to confide in those you trust, like close friends and family members. The most important factor when seeking help isn't necessarily finding someone who knows a lot about your illness, but finding someone who is willing to listen when you're depressed, angry, frustrated - or even just plain old happy. Noticing the emotions you have, accepting them as a natural part of what you're going through, and expressing or sharing your emotions in a way that feels comfortable can help you feel better about things.

The Power of Perception

Several scientific studies have shed light on what many have long suspected: a positive outlook helps people living with chronic illnesses adapt — both physically and psychologically — more quickly and effectively. This 'positiveness' has less to do with a cheery disposition and more about:

- **control** or a sense of having power over personal experience;
- **commitment** or a strong involvement in life's activities; and
- **challenge** — seeing change as a chance to overcome, learn and grow.

3

Possessing a strong sense of these three elements affects your attitude and how you adjust to handling the challenges of a chronic ailment. Of course, one of the first and most important things to accept is the roller coaster of emotions you and those close to you will experience. No matter how positive your attitude, there will be moments of darkness and emotional upheaval. Though each person is unique and responds to chronic illness differently, the suggestions below can enhance your sense of control, commitment and challenge.

Ask for help. While it may seem obvious, it's often difficult to reach out to others for assistance. You may see it as a sign of weakness or worry about being a 'burden' to those around. But ask yourself this: if a loved one was ill and needed your help would you hesitate? Many loved ones may want to assist you, but don't know how. Voicing your specific needs could help you get physical, social and psychological support and lets friends and relatives — who may feel powerless to help — know exactly what they can do to lend a hand.

Communicate. The diagnosis may leave you feeling like you've been stranded on a desert island with no one to truly understand your situation. Resist the temptation to pull away and instead express your feelings and ideas to others. Talking about your emotions — your fears, anger, hopes, dreams etc. — with loved ones can help you unleash bottled up stress and let friends and relatives better understand your emotional journey. Open communication encourages loved ones to share in your thought process and express their own feelings, which they may be afraid or uncomfortable to voice.

Focus on what you can do. Instead of fixating on the things you're no longer physically able to do (e.g. "I can't run or do sports with my son anymore") try to phrase your thoughts to highlight your abilities ("I can still go for short walks and spend time with my son"). Concentrating on your strengths and using positive self-talk will help you maintain your self-confidence during a time when it's needed most.

Explore new ways to relieve tension. If physical limitations mean you can no longer enjoy certain hobbies or physical activities then why not challenge yourself by trying something new? From painting to poetry, swimming to meditation; seize the opportunity to discover a new passion. Whether physical or creative, delving into new pursuits provides an outlet for stress and gives you the chance to learn and grow.

Seek out support. Meet for coffee with a close friend, join a support group of people living with the same illness or turn to a counselor for information and coping strategies. Creating a network of support helps you handle the ups and downs of living with your ailment and reinforces that you're not alone.

Arm yourself with knowledge. Seek out as much information as you can about your condition from doctors, medical journals, reputable Web sites and support groups. The better you understand your condition, the more in control you'll feel coping with the illness and discussing treatment options with your physician. Knowledge is, truly, power.

Play an active role in your health care. The best way to learn about your condition and put yourself in control is to ask questions. There's usually a lot of information to absorb when visiting a doctor. You may need to go over specifics more than once or ask a doctor or nurse to repeat things to be sure you understand everything. This may sound basic, but lots of people hesitate to say, "Hey, can you say that again?" because they don't want to sound stupid. But it takes doctors years of medical school and practice to learn the information they're passing on to you in one office visit!

If you've just been diagnosed with a particular condition, you may want to write down some questions to ask your healthcare provider. For example, some of the things you might want to know are:

- How will this condition affect me?

- What kind of treatment is involved?

- Will it be painful?

- How many treatments will I get?

- Will I miss any school?

- Will I be able to play sports, play a musical instrument, try out for the school play, or participate in other activities I love?

- What can I expect - will my condition be cured? Will my symptoms go away?

- What are the side effects of the treatments and how long will they last?

- Will these treatments make me sleepy, grumpy, or weak?

- What happens if I miss a treatment or forget to take my medicine?

- What if the treatments don't work?

Even though your healthcare provider can't exactly predict how you'll respond to treatment because it varies greatly from one person to the next, knowing how some people react may help you prepare yourself mentally, emotionally, and physically. The more you learn about your illness, the more you'll understand about your treatments, your emotions, and the best ways to create a healthy lifestyle based on your individual needs.

Live in the here and now. While it's natural to long for the healthier days of the past, living there can interfere with your ability to adapt to your condition and impede your enjoyment of today. Savor daily delights — whether it's a nice chat with a neighbor, a compliment from a colleague or a homemade card from your child or extended family.

Keep things in perspective. It's easy for a health condition to become the main focus of a person's life - especially when they first learn about and start dealing with the condition. Many people find that reminding themselves that their condition is only a part of who they are can help put things back in perspective. Keeping up with friends, favorite activities, and everyday things helps a lot.

Set realistic goals. Track progress and create a sense of accomplishment by creating both short and long-term goals for yourself. Make sure, however, these markers acknowledge the physical and emotional realities of your condition. Perhaps you're not able to run a marathon, but you can participate in a 5km walk or you can't possibly finish the 20 things on your list today, but you can get through five. Break goals down into manageable chunks and set yourself up for success by ensuring they're attainable.

Explore your spiritual side. Whether you attend religious services, belong to a meditation group, or volunteer to help those less fortunate, a number of studies suggest that finding greater meaning and purpose in your life can help you physically and emotionally adapt to the challenges of living with a chronic illness.

Loss of control, over your body and its ability, is a major hurdle of living with a chronic illness. But while you may not have control over the way the illness develops and affects your body, you do have control over how you handle these challenges. Learning to cope is an ongoing effort; one that recognizes the difficulties of living with a chronic illness, adjusts to new circumstances, and celebrates the successes along the way.

Supporting a Loved One with a Chronic Illness

Study after study reveals that people who have a chronic ailment adjust better when they have a network of people behind them. All the more reason why it's important for family and friends to focus on their own needs, as well as those of their loved one. Here are a few tips that may help:

Reflecting and adapting. Take a time out for some introspection and openly assess your reaction. Is your natural instinct to dive in and help, retreat or feel angry with the medical staff? Identify and acknowledge these responses, even if they're irrational. You may try to avoid your loved one, for example, because you feel helpless to change the situation. The more honest you are with yourself, the easier it will be to respond in the most supportive, helpful way possible.

Bridging the gap with communication. Don't know what to say or do around your loved one? Tell them. Creating a façade that everything is 'normal' blocks the lines of communication and may unintentionally belittle your loved one's experience. Tell your friend or relative you're worried or that you're not sure how to help, and use open communication. He or she will appreciate your honesty.

Putting yourself in your loved one's shoes. Helping and supporting a chronically ill friend or relative can, at times, become trying. Stress, discomfort or pain brought on by the ailment may make your loved one irritable or unresponsive. When these moments arise, remain patient and calm and remind yourself of the taxing struggle (both physically and mentally) your loved one is enduring.

Acknowledging successes and challenges. Accept there will be good days and bad, whether it's overcoming a round of treatment, taking a step towards rehabilitation, or sharing in their grief when stumbling blocks arise.

Finding support. Just because your friend or relative has the condition, doesn't mean you have to go it alone. Share your fears, stresses and triumphs and connect with others in similar situations. Join an organization that provides support and information to caregivers or family members of people affected by a chronic illness. Some groups deal with chronic illness in general, while others address the concerns and challenges of specific conditions.

And remember: your presence and support are the most meaningful gifts you can give to a loved one living with a chronic illness.

FROM PATIENT TO PROVIDER....

———— ⚜ ————

*"This is not the end, this is not even the beginning of the end.
This is just the end of a beginning." -- Winston Churchill*

Transplantation is not the end of your life; it is not even the beginning of the end. Transplantation is the end of a new beginning. A transplant begins a whole new opportunity at life.

It was January 1992, and I was 18 years old. I had the world by the tail. I had graduated earlier the previous summer in the top of my senior class and was on scholarship to Union University. The grades were in from my first semester in college and were not exactly what I had foreseen myself accomplishing. However, I really did not think a lot of it. I had pledged to a social fraternity and was working nights at Dominoes pizza. I was spending most of my time socializing and less time studying, yet, during the semester I was making the grade. I just did not understand exactly what had happened. I had not felt good for quite some time, but I had just resolved myself to the fact that maybe I was not as prepared for this "college pace" as I thought I was.

My mother had been begging me to go to the doctor for some time. But, being an 18-year-old male, I felt I was invincible and my parents could not tell me anything I did not know already. Don't get me wrong. I was never a rebellious kind of guy. But, like all young males, I thought I could beat the world. The last thing on my mind was a major illness.

What I did not tell my parents was the persistent symptoms I kept ignoring. For a long time now I had noticed myself getting weaker and weaker. I just could not seem to keep up anymore. I slept all the time. Sometimes my hands would cramp and spasm so bad I would have to pull my hands open with my mouth. There were many other things in retrospect I should have had checked out.

One cold Sunday in January, I had gone to church with my parents. My father is the pastor of the church that I attend, and while he was up preaching I was in the back sleeping. I remember one of my friends waking me after church. We went to eat I got sick in the bathroom, but didn't tell anyone. Went home and I went back to sleep. My parents woke me for church Sunday night. Again, while he preached I went to his office and slept. That night, after church during our supper at home, I became violently sick with nausea and vomiting. My mother had had enough. She had my dad pick me up and place me in the car and off to the ER we went.

January 5th, 1992 I entered Jackson-Madison County Emergency Department. I finally volunteered a one-month history of nausea and vomiting. The ER physician evaluated me: vital signs, CBC, UA, and physical exam. I, of course, was ready to go home. Insisting I was just fine, maybe I had a "stomach virus."

The CBC was quite remarkable with a marked anemia with hemoglobin of 5.9 and hematocrit of 17.4. RBC count was 1.97, normal platelets and normal WBC. I remember being very upset and angry when my parents told me they were going to admit me to the hospital. I did not know at that point why. I remember getting in the wheelchair for the ride up to the floor. While on the elevator the orderly looked at my paperwork and said "9W Oncology, at 18 that's rough." I distinctly remember looking up at my mother and asking, "Mom, what's oncology?" I will never forget the look on her face as she looked at me, as only a mother could, and with tears in her eyes she said, "It's cancer."

This was my first "real" encounter with health care. At this point I was terrified. "Oh God help me! I am 18 and have cancer, but this can't be happening to me. I have my whole life ahead of me." This was my first encounter with nursing. I will never forget her name and face. Linda, the angel God sent to me that night. She was calm and gentle. She was reassuring and in control. I will never forget the first sentence she spoke to me nearly 12 years ago. "My name is Linda, I will be your nurse tonight. When I heard we were getting an 18 year old to the floor tonight, my fellow nurses and I got together and prayed for you." She was the ministering spirit sent to me that night! I felt her confidence and it reassured me. I relented and don't remember much that happened after that—I slipped into unconsciousness.

My next memory is also of Linda. I awoke to a discussion between my parents and my physician. There was talk of a bone marrow biopsy. I was terrified! My grandfather had leukemia and I remember him talking about the pain involved in this procedure. But, then there was Linda. She came in, sat down and took my hand and explained the procedure and said, "that she would be there through the whole thing." And she was. She was there holding my hand through the biopsy, and again her caring confidence soothed and comforted me. She was my guardian angel.

A battery of tests was performed over the next few days, as I would drift in and out. My recollections are fuzzy. A blood chemistry taken in the first 24 hours revealed: sodium 134, potassium 4.4, chloride 98, carbon dioxide 13, anion gap 27.4, glucose 148, BUN 153, creatinine 22.9, calcium 4.6, and phosphorus of 12.4! It was determined I suffered from End Stage Renal Disease. A diagnosis I was more than happy to accept. I feel I received my miracle! Not only did God send an angel to my bedside, but he also listened to our request for mercy. Linda has since gone to meet Him and I am sure she is occupying her special place in heaven. There has to be a special place in heaven for nurses like her.

Now I was faced with a new diagnosis and transferred to a new floor. 6N was the renal floor with the dialysis unit. I could not take my angel with me, for she had to stay behind to fulfill the next mission sent her way. However, the Bible is very firm in the fact that God will never "leave us or forsake us." I met my next angel upon admission to that floor. His name was Andy. Andy had the same confidence and control as Linda. I was relieved. He was there to lighten my night and reassure me that "everything was going to be OK." He was a different type of ministering spirit, but he fulfilled his role in that he was exactly what I needed at that point in my crisis, a voice of rationalization and reason. He was a "friend" to be there to explain exactly what the doctor said in language I could understand, a "brother."

Of, course with the evolution of my diagnosis came the inevitable dialysis treatments. They wanted to start them immediately. Andy was there to help me through the emotions. They wanted to put in a Vas-Cath. I was terrified. Andy promised me he would stay through the procedure, and he did. They gave me something to relax. The last thing I remember was Andy telling me everything was going to be all right. He held my hand as I slipped into a sedated state.

I awoke on a dialysis machine and looked into the face of my next angel, Ruth. "Ruthy," as I lovingly came to call her, greeted me with a cheerfulness and compassion of a grandmother. She was an older nurse that had done about "every kind of nursing there's been!" There were many other nurses that took care of me during those dark, early days. Most were really good and I owe my life to the "heath care team" that participated in my care. But, I feel neither my family nor I would have made it through had it not been for 3 angels sent from God: Linda, Andy, and "Ruthy."

To make a long story short after months on dialysis, four hours a day every other day just to stay alive, we started a kidney transplant work-up. My Uncle Gary was the transplant coordinator at The University of Tennessee in Memphis and had been for years. There were many holiday family dinners and visits that had been interrupted by his pager, and he would have to run off at all hours of the day and night to "pick-up" organs to give people their life back. He was one of my heroes. We never expected it to hit so close to home. Never the less, my mother was a perfect match and did not hesitate at the chance to give me the "gift of life" for the second time in my life, at birth and now.

Again, just when I needed reassurance and care in this tumultuous time in my life, there were angels in uniform to help me through. Dr. A. Osama Gaber and Dr. Santiago Vera at UT along with the nursing staff on the transplant floor never missed a beat. With skill and professionalism they "cared" for my mother and me. They were God's servants here on earth. Then it began to dawn on me why I had to go through this trial in my life. I knew what I was supposed to do once I got my life back. I could become an angel for someone else. I knew then I had to become a healthcare professional.

Once back at Union, now status-post transplant, I began to pursue my "mission." I accomplished the goal I set to achieve. First with my Associate degree and then my Bachelor's, I believe I have been fulfilling my "mission" on earth. From my early years in the Cardiac Care Unit at Jackson Madison County General Hospital, to my position as Physician Extender for a Cardiology Group, to my current "ministry" as a Family Nurse Practitioner. I have been Linda, I have looked through Andy's eyes, and I have cared as "Ruthy." I have held the hands of the dying as they have passed from this life to the next. I have cried with their families and attended their funerals. I have rejoiced with patients who have won their battles with disease and rejoiced with them and their families as I see them for their yearly check-ups.

From 1995-1997 I worked as a "RN team builder" for the Jackson-Madison County Hospital Cardiac Care Unit. During my time as a team builder I rotated through all areas of the hospital with CCU as my home specialty. I have had the chance to work in the Medical, Surgical, Cardiovascular Surgical, and Neurological Intensive Care Units. I also had the opportunity during that time to work both Cardiac Progressive Care and Neurological step-down units. Over this period of two years I also worked as Critical Care Coordinator, managing five Intensive Care Units, as well as, five step-down units. Part of my responsibilities as Critical Care Coordinator was to "scrub-in" on various procedures in the Intensive Care Special Procedures Suite.

While working for Jackson General I was also involved in nurse education and worked on my days off as an Advance Cardiac Life Support (ACLS) instructor.

In 1997, a cardiologist offered me a position as his "physician extender," a title that Jackson-General Hospital has given to non-physician staff that assists physicians in various positions in the hospital. This was a tremendous opportunity for me and the experience was invaluable. For over 6 years, I was given multiple responsibilities and have earned the complete, unquestioning trust of my employing physician. My multi-tasking included the treatment of patients in the acute setting (i.e., ER evaluation and treatment, and hospital rounds), as well as, seeing patients for follow-up in the clinic setting (i.e., stress testing and clinic visits). I also assisted with procedures in the hospital (i.e., CVP placement, Swan Ganz insertion, personally removed balloon pumps, etc.).

In 1999, the position of Research Coordinator/Liaison was added to my "platter." My physician and I started a now nationally recognized research program. It was my responsibility to coordinate and run both hospital and office based research studies along with my duties as described above. This was a new and interesting challenge over the years. However, hard work rewarded us with national recognition and multiple publications. I have personally been responsible for over 10 studies, two of which have been published in the LANCET a world acclaimed medical journal. One study in particular, ESPRIT, I ran the number one site in the country. I have also been co-author on several publications in other medical journals. I had one abstract accepted for presentation at the American College of Cardiology,

2002 and two abstracts were published in the American Journal Of Cardiology. A case study analysis I researched was presented by my physician at the Transcatheter and Therapeutics conference in Washington D.C., 2000.

I have, in my spare time, worked as a consultant for pharmaceutical companies involved with nurse education and giving lectures on Acute Coronary Syndrome and the use of Glycoprotein IIb/IIIa inhibitors as well as Acute Myocardial Infarction and TNKase, a newly developed thrombolytic agent. I also give lectures on Congestive Heart Failure and new treatment options of which I have been involved in the research.

I graduated the University of Tennessee Health Science Center (UTHSC) in December of 2003. I received my degree as a Family Nurse Practitioner. I was the President of my class and had many honors while at UT. I sat on the SGA executive committee, I was inducted into Sigma Theta Tau and the Imhotep Society, leadership and scholarship honor societies, I was on an Advisory subcommittee for the search for a new chancellor for the UTHSC, as well as, various other student government committees.

I am currently seeking my Doctorate in Naturopathic Medicine from Clayton College. I am practicing Primary Care medicine full-time in Henderson, TN. I am seeing all patients ranging from pediatric to geriatrics. I am very involved in my church, as well as politics. I have had the pleasure of meeting with many Congressional leaders in both the House of Representative and the Senate. I attended the Inaguration Ceremony of our 43rd President of the United States George W. Bush both times.

I was most recently inducted into the Who's Who among Executive and Professionals in the United States. I have been married for 10 years to the love of my life, Holly, and we had our first child, Jeffrey Walter Young, III (TREY) in April 2004.

I say all of that to say this…transplantation is not the end of your life, but a chance at a new beginning.

From patient to provider, this is why I do what I do. This is my ministry, my chance to make a difference in the life of other people, a chance to be the hands, face, and voice of Jesus. I knew I could make even more of an impact on my community and my patients. I felt I owed this to all those "angels" that cared for me.

One day I was walking through the hospital on my rounds. I saw a familiar figure in the distance. As I got closer, I noticed who it was, my angel "Ruthy." She had retired shortly after I left the hospital. I had not seen her in years! I walked up to her and said, "Ms. Ruthy?" She looked at me puzzled. I said "Jeff Young, dialysis 12 years ago." Her face softened, she smiled, and we embraced. Then I knew furthering my education was the right thing for me to do. Every caregiver eventually becomes a patient and their perspective changes. I have had the unique chance to go from critically ill to caregiver, patient to provider. Linda is smiling. I know that I was given a second chance at life for a reason. And though I have been somewhat successful, I am obligated to do more. I will be held accountable for what I have done with my second chance. When I stand before God one day I want to be able to look into His face and say, "I made a difference in the lives of others."

NEVER GIVE UP!!!

WHAT IS KIDNEY FAILURE?

⎯⎯⎯⎯⎯ ⚜ ⎯⎯⎯⎯⎯

To understand kidney failure we must start at the basics. What are the kidneys and what do they do? To quote my old anatomy and physiology professor, "the structure and function." Kidneys are the two bean-shaped organs that filter wastes from the blood and form urine. The kidneys are located near the middle of the back. They send urine to the bladder.

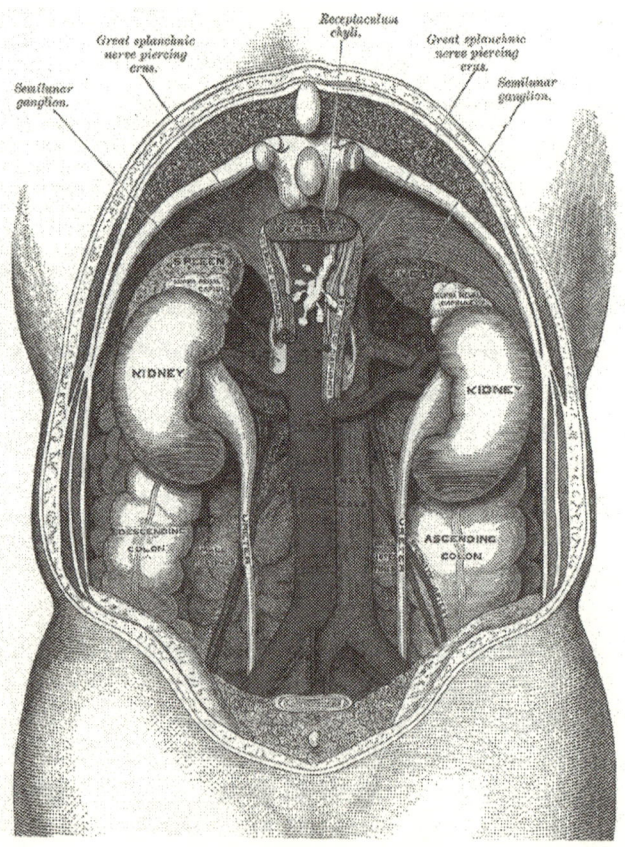

Kidney failure is a condition in which the kidneys are less able than normal to perform their functions of removing toxic wastes, removing excess water, helping to control blood pressure, helping to control red blood cell manufacture and helping to keep the bones strong and healthy. Kidney failure can be acute or chronic. Advanced chronic kidney failure is called end-stage renal failure (ESRF). A diagnosis of ESRF is irreversible failure of both kidneys that necessitates treatment by regular dialysis or kidney transplant.

What causes kidney failure?

There are many causes of kidney failure. These include: Diabetes - high levels of blood glucose (blood sugar) can damage the kidneys; High blood pressure - this is a cause of kidney failure in itself, but can also occur because of existing kidney damage and, if left untreated, leads to further damage and progression of kidney failure; Nephritis - this is when there is inflammation of the glomeruli , which are part of the filtration unit of the kidneys; Pyelonephritis - this is when there is inflammation around the draining system of the kidney; Renovascular disease - The arteries (blood vessels) from the heart to the kidneys become diseased and fail to provide the kidneys with an adequate blood supply; Polycystic kidney disease (PCKD) - this is a disease in which both kidneys become filled with cysts; Analgesic nephropathy and drugs - misuse or overuse of certain painkillers and drugs can damage the kidneys; and then unknown causes - In about 30% of cases, the cause of the kidney failure is never discovered.

Kidney failure itself is not hereditary, but some of the diseases that cause kidney failure are. Polycystic kidney disease (PCKD) is one of the most common of these. The other genetic disorders that can lead to kidney failure include a condition called Alport's Syndrome, but are very rare. Diabetes, one of the leading causes of kidney failure, does run in families. Let us now cover these causes in a little more detail

Nephritis

The term nephritis covers a group of conditions in which there is inflammation within the kidneys ("neph-" means "kidney" and "-itis" means "inflammation"). Sometimes this condition is described, more specifically, as glomerulonephritis or GN ("glomerulo" refers to the glomeruli, which are part of the kidneys' filtration unit). GN can be caused by Systemic Lupus Erythematosus (SLE). SLE is an autoimmune disease, which affects small blood vessels throughout the body - in particular the glomeruli in the kidneys. The causes of most types of nephriitis are not known.

Pyelonephritis

"Pyelo" refers to the draining system (tubules) of the kidneys (they look like a funnel) and nephritis means "kidney inflammation". So, pyelonephritis means "inflammation of the kidney draining system".

Pyelonephritis can sometimes be linked to repeated kidney infections. These may have gone undetected for many years, perhaps having occurred in childhood. It is sometimes caused by a condition called "reflux" or "reflux nephropathy". This is when a faulty valve where the ureter enters the bladder allows urine to flow back up the ureter, causing problems in the kidney.

Polycystic kidney disease (PCKD)

This is an inherited disease (a disease which runs in families) in which both kidneys become filled with cysts (abnormal pouches of fluid). If someone has PCKD, they have a 50% chance of passing the problem on to their children.

Polycystic kidneys, although often large because of the cysts, do not work well and most people with PCKD eventually develop kidney failure. The cysts in PCKD can remain a problem after treatment for kidney failure has started. A cyst can burst, bleed or become infected - any of which may cause pain. Occasionally a large cyst may need to be drained through a long hollow needle. In some patients before transplantation, one or both kidneys may need to be removed.

Renovascular disease

'Reno' refers to the kidneys and 'vascular' refers to blood vessels. A narrowing of the arteries, which supply blood to the kidneys, causes Renovascular disease. The narrowing of the arteries is caused by the inside of the artery becoming "furred up" with cholesterol and other fats. The kidneys become damaged from not receiving an adequate blood supply.

Analgesic nephropathy

Analgesics are medicines intended to relieve pain, such as aspirin and ibuprofen. These drugs are not dangerous for most people as long as they are taken according to the recommended dosage. However, if people take painkillers every day over a long period of time, take more than the recommended dose, or regularly take a combination of painkillers, their kidneys may become damaged.

If you are reading this book you probably have been diagnosed with one of these condition or a loved one has. Also, if you are reading this you have been discussing the possibility of kidney transplantation with you healthcare provider. So let's hopefully answer some of the questions you might have.

What is a kidney transplant?

When a kidney transplant takes place, a kidney from another person is put into the body of someone with kidney failure. The new kidney can then do the work of the failed kidneys so the patient no longer needs dialysis.

Who does the kidney come from?

Kidneys come from one of two types of donor: a living relative of the person who is receiving the kidney or someone who has died (a 'cadaveric donor').

A cadaveric donor is someone who has died in an accident or from a brain hemorrhage and who has been on a life support machine. Organs can only be donated once the person is 'clinically dead' which means he or she would not survive were the life support machine to be switched off. Allowing a relative's organs to be used for donation usually brings some comfort to the bereaved family.

A living related donor will have only one kidney after the operation but is still able to lead a full and active life.

Who is suitable for a kidney transplant?

Most people with kidney failure are suitable for a kidney transplant. The only exceptions are people with serious heart failure or cancer. Each patient is tested for certain viruses before going on to the waiting list. Having a virus like HIV or hepatitis will not prevent your child having a transplant, but the doctors will need to be more careful. If you refuse to let your child be tested for these viruses, it will not be possible for him or her to go on to the waiting list.

Are there any risks or side effects with a transplant?

A transplant gives most people with kidney failure the best quality of life possible and offers an end to dialysis and special diets. However, not all transplants are successful. If the transplant is successful, your child will need to take certain drugs for the rest of his or her life to prevent the kidney being 'rejected'. These drugs have side effects, but there may be ways of minimizing them. Ask your doctor, nurse or pharmacist for details.

Finding the right kidney

For a transplant to be successful, the tissues of the patient must be reasonably similar to those of the donor. This is called a 'good match'. In order to find out whether a kidney will be a good match or not, your child will need to have various tests to find out his or her blood and tissue type.

Blood groups and tissue types

Everyone has a blood group, which remains the same throughout life. It is based on which of two types of protein (antigens) are present in your body. The main blood groups are A, B, AB and O. People with blood group A have only antigen A in their body, people with blood group B have antigen B, people with blood group AB have both antigens, and people with blood group O have neither. The most common type is O, followed by A.

Everyone also has a tissue type, which does not change. It is based on the antigens present on the surface of most cells. You have only one tissue type but it is made up of six different parts. The three main sorts of tissue type are A, B and DR. There are about 20 different versions of the three tissue types. The more tissue types you match with the donor the less likely the body is to reject the kidney.

Preparing for transplant

Before your child joins the transplant waiting list, you will meet your nephrologist (a doctor who specializes in kidney disease) and your clinical nurse specialist. They will explain all about the kidney transplant and what to expect. It is also likely that you will meet the renal social worker ("renal" means anything to do with the kidneys) and a psychologist to make sure your child and your family is ready and prepared mentally for joining the list.

The play specialist on the ward will help you to prepare your child for transplant. He or she will show your child a photo book of children who have already had a transplant.

The transplant waiting list

If a person is a suitable candidate for a kidney transplant, their name is put on a waiting list for an organ. Unfortunately, there are many more people on the waiting list than there are organs available each year. There are now more than 51,000 people in the U.S. waiting for a donor kidney. Waiting time may extend several years.

Kidneys are allocated based on a combination of the length of time a person has been on the wait list and how well matched the kidney and the recipient is to each other.

The ideal donor is a living family member, or other unrelated individual who is emotionally close to the patient. This reduces the waiting time, the procedure can be planned in advance, and the results are better than transplantation using a kidney from a deceased donor.

When a suitable kidney is found

People on the waiting list should be contactable at any time of the day or night. Sometimes, hospitals give out bleeps to be used only for this purpose. When you are contacted, you will need to travel to your Transplant Center right away. If traffic could be a problem, the transplant nurse could ask the police to escort you. It is important to get to the hospital as quickly as possible so that blood tests can be carried out.

Does the operation have any risks?

Every anesthetic carries a risk of complications but this is very small. Your child's anesthetist is an experienced doctor who is trained to deal with any problems. After an anesthetic, some children feel sick and vomit, have a headache, a sore throat or feel dizzy. These effects are usually short-lived. Any surgery carries a small risk of infection or bleeding. Your child will be given a course of antibiotics to reduce the chance of infection.

What does the operation involve?

The operation lasts about four hours and will be carried out under a general anesthetic. The surgeon will make an incision (cut) on your child's abdomen either on the right or left side depending on where he wants to insert the kidney. He will not usually remove your child's failed kidneys unless they are causing problems. The new kidney will be placed above the pelvic brim, which is the part of the bone that sticks out in the front above hip level. The surgeon will connect up the various tubes of the new kidney with your child's arteries and veins so that the kidney is supplied with blood. He will also connect the drainage tube (ureter) to your child's bladder so that urine can drain away. A stent (plastic tube) may be inserted to stop the ureter blocking. At the end of the operation, the surgeon will sew up the incision.

What happens afterwards?

Following kidney transplant surgery, the recipient will remain in the hospital for a few days to a week. Walking is encouraged as soon as the anesthesia wears off. Generally, he or she will also begin eating within 24 hours after surgery. A bladder catheter inserted during surgery can be removed in a few days.

Recovery for the living donor is usually speedy, and they may return home in 2 to 5 days. They too will awake with a bladder catheter, which is removed after the first day. As soon as the anesthesia wears off, the donor is usually up and around and can resume eating.

Because the organ will be identified as foreign by the recipient's immune system, rejection of the new kidney is always a possibility. Powerful drugs called immunosuppressants are given starting at the time of kidney transplantation surgery to try to prevent rejection. Observation during the first weeks following surgery is used to prescribe accurate immunosuppressant drug dosage.

Prior to discharge, the transplant team reviews information with the patient, gives instructions for follow-up care and medications, and answers the patient's questions. A prescribed rehabilitation program for the recipient will continue at home including exercise, nutrition, and the continuation of immunosuppression and other medications. The signs of rejection are also discussed with the transplant recipient and family.

Living donors do not have to take any specific medications or maintain a special diet as a result of kidney donation.

Returning home

At-home rehabilitation for kidney transplantation depends on the individual. The transplant team will give specific instructions. In general, walking is recommended to restore strength, but heavy lifting should be avoided four to six weeks following transplant surgery. Other activities, such as driving may usually begin when the incision is free of pain. Sexual activity can resume when one is comfortable. A desire to become pregnant should be discussed ahead of time with the transplant team to determine if and when this is recommended.

At-home recovery is similar for the donor. The donor generally only needs to return for a postoperative check-up.

Continued follow-up visits are required for check-ups for the recipient. These begin soon after returning home. Initially, outpatient visits may occur weekly or even more often for the recipient, and as time progresses the frequency of follow-up visits usually decreases.

Possible post-operative complications may arise following kidney transplant surgery. They include:

- Leaking or blocked ureter
- Bleeding

- Lymphocele (fluid collection around kidney requiring drainage)
- Problems with blood vessels
- Transplant renal artery stenosis

Leaking or blocked ureter - During transplant surgery, the ureter from the donor kidney is connected to the recipient's bladder. Sometimes the connection can leak. The treatment may involve inserting a very thin tube, called a stent. This provides the necessary scaffolding needed for the tissues to heal. Sometimes another operation is required.

Major bleeding - It is uncommon for a kidney transplant patient to experience bleeding after surgery. There may be leakage from the newly connected blood vessels and other operated surfaces. Patients may need a blood transfusion.

Problems with blood vessels - In general, complications can arise with blood vessel connections between the donor kidney and the recipient's blood vessels. A more serious complication is a clot in an artery or vein attached to the kidney. If a clot occurs and blood is unable to flow to and from the kidney, the kidney may cease to function.

Transplant renal artery stenosis - This is when the artery going to the kidney narrows, limiting blood flow to the kidney. This can also make it difficult to keep blood pressure under control. Treatment typically involves expanding the narrowed segment using a small balloon.

Other problems include the long-term risks of immnosuppression. These include complications related to too much or too little immunosuppression:

- Rejection
- Cancer
- Infection

Rejection - It is fairly common for a transplant patient to experience rejection episodes. The body identifies the new organ as foreign and may try to reject it. The immunosuppressive medications prevent rejection in 50 to 75% of cases. Changes may be made in the medications including an increase in dosage or the use of additional drugs to stop the rejection. Some episodes can cause permanent damage to the new kidney. This may reduce longevity of the organ. Rejection occurs when the body recognizes the new kidney as 'foreign' and the immune system starts to attack it. This can happen even when the new kidney is a 'good match' for the patient and most people suffer some level of rejection at some point. It is the most common cause of kidney transplant failure.

The symptoms of rejection can include:

- temperature
- pain or tenderness around the kidney
- blood in the urine
- a smaller amount of urine than usual
- general lethargy or malaise

If you see any of these signs, contact the Transplant Unit immediately.

Rejection can either be acute (sudden) or chronic (ongoing).

Acute rejection: This can sometimes cause pain and fever, but not everyone has symptoms during rejection. The doctors will need to carry out some tests to see if the kidney is still working. These will include a renal ultrasound and scans to check the blood supply to the kidney. To confirm a diagnosis of acute rejection, the doctors may need to take a kidney biopsy. If the diagnosis is confirmed, your child will need a course of high-dose steroids. This may stop the rejection process but sometimes other drugs will be needed. These are likely to be given intravenously and in most cases they will stop the rejection. These drugs can have nasty side effects, but the ward staff will try to minimize these as far as possible.

Chronic rejection: This term implies that the body is rejecting the kidney whereas it is more correct to say that the kidney is wearing out. If the kidney stops working, your child will need to re-start dialysis and possibly wait for another transplant.

Cancer - Studies show that an estimated 6% to 8% of transplant patients will develop cancer over their lifetime with the transplant. This risk is higher than in the general population. Skin cancer is the most common, and is typically treated successfully. Some cancers result from the effects of the immunosuppressive medications and others are common cancers that occur at a higher rate in immunosuppressed individuals.

Infection - The immunosuppressant medications increase the risk of less serious and common infections such as urinary tract infection. In addition, they are associated with more serious infections like pneumonia. Finally, uncommon infections that do not affect non-immunosuppressed persons can occur.

WHAT IS HEART FAILURE?

———————— ⚜ ————————

What exactly is the heart? The heart is one of the major organs of the body. Simply put---something you can't live without! Basically the heart is a muscle that pumps blood through the body. It is about the size of your fist and is located in the left center of your chest. It pumps about 1800 gallons of blood to all areas of the body 24/7/365 through a complex series of "pipes" called arteries and veins. This is just a very complex plumbing system. I have always described, to my patients, the heart is like a house. As your house ages you can begin to have problems. Think about it. If you have owned a home for very long, or rent, or whatever, as times goes by you can develop various problems. You can have electrical problems, plumbing problems, and/or the doors don't work like the used too. Not to oversimplify things, but the heart is the same thing. Your heart is the house. The house has plumbing/pipes your doctor calls these arteries and veins. The house has electric current that runs through it, your doctor refers to this as SA node, AV node, Bundle of HIS, bundle branches, etc. The electricity causes the pump/heart to work/squeeze. Your house has rooms; your doctor likes to refer to these as the four "chambers of the heart" right atrium, right ventricle, left atrium, and left ventricle. If your house has rooms it must have doors, your doctor likes to refer to these as heart valves (tricuspid, pulmonary, mitral, and aortic).

Heart failure, also called congestive heart failure, is a disorder in which the heart loses its ability to pump blood efficiently. Heart failure is almost always a chronic, long-term condition, although it can sometimes develop suddenly. This condition may affect the right side, the left side, or both sides of the heart.

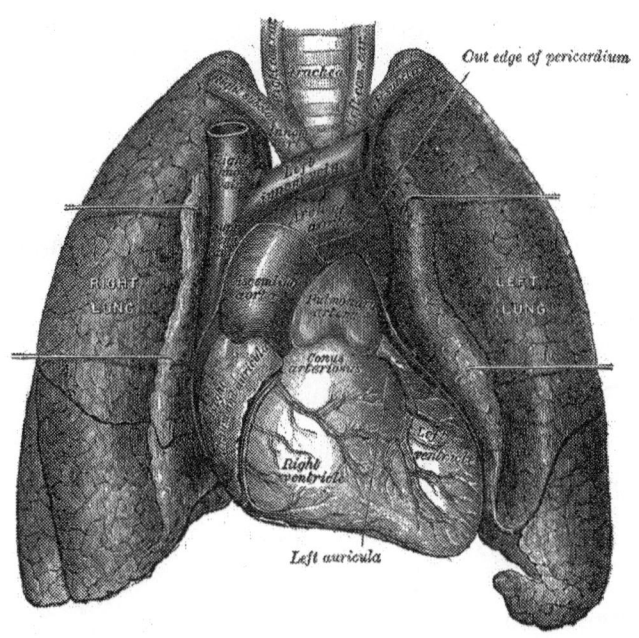

As the heart's pumping action is lost, blood may back up into other areas of the body:

- The liver
- The gastrointestinal tract and extremities (right-sided heart failure)
- The lungs (left-sided heart failure)

With heart failure, many organs don't receive enough oxygen and nutrients, which damages them and reduces their ability to function properly. Most areas of the body can be affected when both sides of the heart fail.

The most common causes of heart failure are hypertension (high blood pressure) and coronary artery disease (for example, you have had a heart attack). Other structural or functional causes of heart failure include the following:

- Valvular heart disease
- Congenital heart disease

- Dilated cardiomyopathy
- Heart tumor

Valvular heart disease:

There are four valves that control the flow of blood through the four chambers of the heart (listed below in the same order as blood flows through the heart):

- Tricuspid valve (in the right side of the heart between the right atrium and right ventricle)
- Pulmonic valve (in the right side of the heart between the right ventricle and the pulmonary artery)
- Mitral valve (in the left side of the heart between the left atrium and left ventricle - and the only valve with 2 leaflets instead of 3)
- Aortic valve (in the left side of the heart between the left ventricle and the aorta)

The Tricuspid and Mitral valves open during diastole, which is the period of time when the ventricles are filling with blood. The Pulmonic and Aortic valves open during systole when blood is being ejected from the heart.

Disease can affect these valves in two ways.

- Stenosis is a narrowing of the valve so that blood cannot move through as freely as necessary.
- Regurgitation is a failure of the one-way valve so that blood flows back through the valve in the wrong direction.

The valves most commonly affected by disease are the mitral valve, which controls flow of the blood from the left upper chamber, or atrium, to the left lower chamber, and the aortic valve, which controls blood flow out of the left ventricle to the rest of the body.

Congenital heart disease

Congenital heart disease (CHD) is a broad term that can describe a number of different abnormalities affecting the heart. Congenital heart disease is, by definition, present at birth although its effects may not be obvious immediately. In some cases, such as coarctation of the aorta, it may not present itself for many years and a few lesions such as a small ventricular septal defect (VSD) may never cause any problems and are compatible with normal physical activity and a normal life span.

According to the American Heart Association, approximately 35,000 babies are born each year with some type of congenital heart defect. Congenital heart disease is responsible for more deaths in the first year of life than any other birth defects. Many of these defects need to be followed carefully; though some heal over time, others will require treatment

Some congenital heart diseases can be treated with medication alone, while others require one or more surgeries. The risk of death from congenital heart disease surgery has dropped from approximately 30% in the 1970s to less than 5% in most cases today.

Cardiomyopathy

The causes of cardiomyopathy are multiple and may include nutritional deficiencies, valvular heart disease, anemia, stress, viral infections (rare), alcoholism (alcoholic cardiomyopathy), coronary artery disease, and others. In some cases, no cause can be identified (idiopathic cardiomyopathy).

Enlargement of the heart muscle (dilated cardiomyopathy) is the most common type of cardiomyopathy. Restrictive cardiomyopathy is another type that occurs when there is damage to the heart and scarring (fibrosis) or deposits develop in the heart muscle.

Some risk factors include obesity, having a personal or family history of cardiac disorders (such as mycarditis) and alcoholism.

Dilated cardiomyopathy occurs in approximately 2 out of 100 people. It can affect all ages and both sexes, but is most common in adult men.

Heart Tumor

A myxoma is a primary heart (cardiac) tumor. This means that the initial site of the growth was within the heart, which is uncommon with heart tumors. Most have spread from elsewhere in the body (metastasized).

Primary cardiac tumors are rare, but among them, myxomas are the most common. Over 80% of myxomas occur in the left atrium, usually beginning in the wall that divides the upper chambers of the heart (atrial septum) and growing into the atrium. However, right atrial myxoma may also occur.

Myxomas are more common in women.

Approximately 10% of all myxomas are familial, which means that they were genetically inherited. Familial myxomas tend to produce symptoms at a younger age than non-familial tumors and are often multiple, that is, are present in more than one site within the heart and can include the ventricles (lower heart chambers).

Heart failure becomes more common with advancing age. You are also at increased risk for developing heart failure if you are overweight, have diabetes, smoke cigarettes, abuse alcohol, or use cocaine.

Heart transplants are the fourth most common (corneas, kidneys and liver are the most common) transplant operations in the U.S. (over 2,200 cases per year). A healthy heart is obtained from a donor who has suffered brain death but remains on life-support. The healthy heart is transported in a special solution that preserves the organ.

Successful heart transplants have been conducted on newborn infants, children and adults including people past the age of 60.

The evaluating team considers many factors to decide whether a person should be placed on the waitlist for a transplant. The person's general health and suitability for major surgery are taken into account. Risk factors such as these are considered carefully and may result in a recommendation against transplant surgery:

- Emphysema
- Liver or kidney problems
- Poor leg circulation
- Smoking
- Other conditions that may be treated before the transplant. Treatment of these problems improves the chance of success and reduces complications.

A heart transplant would not be performed for people with certain conditions. These include:

- Most cancers, unless successfully treated at least five years previously
- Infections that cannot be completely treated or cured, such as tuberculosis
- Severe lung, liver, or kidney problems that would make the operation too risky

It is a normal reaction of the body to reject the donated organ. Anti-rejection drugs are prescribed to prevent this rejection. The candidate must be willing to take anti-rejection medicines indefinitely to keep the body from rejecting the donor heart. The person will also need lifelong follow-up by health care professionals.

Waiting

If a person is a suitable candidate for a heart transplant, their name is put on a waitlist for an organ. Unfortunately, there are many more people on the waitlist than there are organs available each year. There are now more than 4,000 people in the U.S. waiting for a donor heart. Waiting time may extend several years.

People waiting for donor hearts are grouped by the severity of illness and other medical factors such as blood type. Within any given severity of illness and blood type group, hearts are allocated based on the length of time a person has been on the waitlist.

The Transplant Surgery

When a donor heart becomes available, time is critical. The heart must be transplanted into the patient receiving the organ within 4 to 5 hours. A team of surgeons and anesthesiologists performs an operation to remove the heart from the donor. Additional surgical teams may be present to remove other organs. After the heart is removed from the donor, it is preserved and packed for transport. Although the donor is brain dead, this procedure is treated like any other operation using standard surgical practices and sterile techniques. Once the operation is complete and the incisions are closed, the donor's body is prepared for funeral or cremation. Organ procurement surgery respects the body and an open casket funeral is possible if desired.

In the meantime, a recipient is located and prepared for surgery as well. Preparation involves administration of general anesthesia, and placement on a cardiopulmonary bypass machine. Because the blood must continue to be oxygenated during the procedure, the cardiopulmonary bypass machine will perform this task until the new heart is transplanted and beating.

Electric shock may be used to start the beating of the newly transplanted heart. When the heart is successfully beating, the surgeons check for any signs of bleeding. Drugs are administered to stabilize the heart rate and blood pressure. The patient will begin recovery in the intensive care unit (ICU).

After Transplant Surgery

Following heart transplant surgery, the patient may remain on an artificial breathing machine for the first 12 to 24 hours of recovery. Depending on progress, some patients are moved out of the ICU in a few days. Generally, he or she will also begin eating within the week following surgery.

Because the organ will be identified as foreign by the recipient's immune system, rejection of the new heart is always a possibility. Powerful drugs called immunosuppressants are given starting at the time of heart transplant surgery to try to prevent rejection. Early signs of rejection do not always cause symptoms. Therefore, tiny biopsies of the heart muscle are taken from all recipients on a regular basis for examination under the microscope. To get the biopsy tissue, a small tube is inserted in the neck and down the jugular vein into the heart. If rejection is detected on the biopsy, treatment consists of additional immunosuppression until the rejection episode is reversed.

Prior to discharge, the transplant team reviews information with the patient, gives instructions for follow-up care and medications, and answers the patient's questions. A prescribed rehabilitation program will continue at home including exercise, nutrition, and the continuation of immunosuppression and other medications. The signs of rejection are also discussed with the patient and family.

Returning Home

At-home rehabilitation is a gradual process and depends on the individual. The transplant team will give specific instructions. In general, walking is recommended to restore strength, but heavy lifting should be avoided for four to six weeks following transplant surgery. Other activities, such as driving may usually begin when the incision is free of pain. Sexual activity can resume when one is comfortable. A desire to become pregnant should be discussed ahead of time with the transplant team to determine if and when this is recommended.

Follow-up visits are required for check-ups and additional heart tissue biopsies (described in Step 4). These begin soon after returning home. Initially, outpatient visits may occur weekly or even more often, and as time progresses the frequency of follow-up visits usually decreases.

Possible post-operative complications may arise following heart transplant surgery. They include:

- Vascular problems (bleeding)
- Arrhythmias (irregular heart rhythm)
- Lung problems (collapsed lung; pneumonia)
- Incision problems

Vascular problems - It is possible for a heart transplant patient to experience bleeding after surgery. There may be leakage from the sutures or oozing from operated surfaces. Reports indicate that this occurs in about 5% of heart transplant patients, and an additional operation may be required within the first 24 to 48 hours after the transplant to resolve the problem.

Irregular heart rhythms may occur following heart transplant surgery. These are usually treated with medication, but may sometimes require placement of a pacemaker.

Other problems include the long-term risks of immunosuppression. These include complications related to too much or too little immunosuppression:

- Rejection
- Cancer
- Infection

Rejection - It is fairly common for a transplant patient to experience rejection episodes. The body's immune system detects the new organ as foreign and may try to reject it. The immunosuppressive medications prevent rejection in 50 to 75% of cases. Changes may be made in the medications including an increase in dosage or the use of additional drugs to stop rejection. Some rejection episodes can cause permanent damage to the new heart. This may result in narrowing of the coronary arteries of the transplanted heart and may reduce longevity of the organ.

Cancer - Studies show that an estimated 6% to 8% of transplant patients will develop cancer over their lifetime with the transplant. This risk is higher than in the general population. Skin cancer is the most common, and it is typically treated successfully. Some cancers result from the effects of the immunosuppressive medications and others are common cancers that occur at a higher rate in immunosuppressed individuals.

Infection - The immunosuppressant medications increase the risk of less serious and common infections such as urinary tract infection. In addition, they are associated with more serious infections like pneumonia. Infection of the sternal incision can be life threatening and difficult to treat. Finally, uncommon infections that do not affect non-immunosuppressed persons can occur.

WHAT IS LIVER FAILURE?

---------- ❧ ----------

To understand liver failure we must start at the basics. What is the liver and what does it do? To quote my old anatomy and physiology professor, "the structure and function." The largest glandular organ in the body shaped like a cone, the liver is a dark reddish-brown organ that weighs about 3 pounds, located in the upper right portion of the abdominal cavity. The liver has many functions that include, but are not limited to the production of protein and cholesterol, the production of bile and clotting factors, the storage of sugar in the form of glycogen, and the breakdown of carbohydrates, fats, and proteins. The liver also breaks down and excretes many medications. Your liver helps fight infections and cleans your blood. It also helps digest food and stores energy for when you need it.

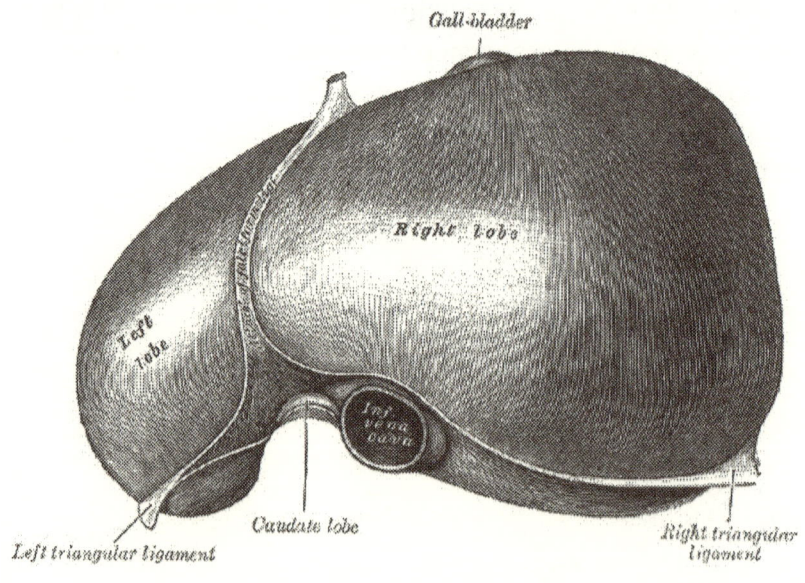

What causes the liver failure?

There are many different causes of liver failure as with any other disease process. The following are the various causes for liver failure:

- Liver damage (cirrhosis or primary biliary cirrhosis)
- Long-term active infection (hepatitis)
- Hepatic vein clot (thrombosis)
- Birth defects of the liver or bile ducts (biliary atresia)
- Metabolic disorders associated with liver failure (Wilson's disease)

Cirrhosis

Cirrhosis is the result of chronic liver disease that causes scarring of the liver (fibrosis - nodular regeneration) and liver dysfunction. This often has many complications, including accumulation of fluid in the abdomen (ascites), bleeding disorders (coagulopathy), increased pressure in the blood vessels (portal hypertension), and confusion or a change in the level of consciousness (hepatic encephalopathy).

Primary Biliary Cirrhosis

Primary biliary cirrhosis is an inflammation (irritation and swelling) of the bile ducts of the liver resulting in narrowing and obstruction of the flow of bile. This obstruction damages liver cells.

Hepatitis

Hepatitis is inflammation of the liver. The disease can be caused by:

- Infections from parasites, bacteria, or viruses (such as Hepatitis A, B, or C)

- Liver damage from alcohol, drugs, or poisonous mushrooms
- An overdose of acetaminophen, which is rare but deadly
- Immune cells in the body attacking the liver and causing autoimmune hepatitis

Hepatitis may start and resolve quickly (acute hepatitis), or cause long-term disease (chronic hepatitis). In some instances, progressive liver damage, liver failure, or even liver cancer may result.

The severity of hepatitis depends on many factors, including the cause of the liver damage and any underlying illnesses you have. Hepatitis A, for example, is generally short-lived, not leading to chronic liver problems.

Common risk factors include:

- Intravenous drug use
- Acetominophen overdose -- the dose needed to cause damage is close to the effective dose, so be careful to take it only as directed; DO NOT use if you already have underlying liver damage
- Risky sexual behaviors (like having multiple sexual partners and unprotected intercourse)
- Eating contaminated foods
- Travel to an endemic area, like Asia, Africa, or South or Central America
- Living in a nursing home or rehabilitation center
- Family member who recently had hepatitis A
- Alcohol use
- AIDS

- Blood transfusion received prior to 1990 (before hepatitis C blood test was available)
- Newborns of mothers with hepatitis B or C (can be transmitted during delivery)
- Healthcare workers, including dentists and dental hygienists, because of blood contact
- Receiving a tattoo

Most common of all hepatitis causes for liver transplant is Hepatitis C (HCV). Hepatitis C infection is caused by hepatitis C virus (HCV). Persons who may be at risk for hepatitis C are those who:

- Received a blood transfusion prior to July 1992
- Received blood, blood products, or solid organs from a donor who has hepatitis C
- Injected street drugs or shared a needle with someone who has hepatitis C
- Have been on long-term kidney dialysis
- Have had frequent workplace contact with blood (for instance, as a healthcare worker)
- Have had sex with multiple partners
- Have had sex with a person who has hepatitis C
- Shared personal items, such as toothbrushes and razors, with someone who has hepatitis C
- Were born to hepatitis C infected mother

The prevalence of hepatitis C infection is approximately 4 million people in the United States or about 1 in 70 to 100 people.

Biliary Atresia

Biliary atresia is an obstruction of the bile ducts caused by their failure to develop normally in the fetus. This is a congenital condition (present at birth). Biliary atresia is caused by the abnormal development of the bile ducts inside or outside the liver. The purpose of the biliary system is to remove waste products from the liver, and to carry bile salts necessary for fat digestion to the small intestine.

In babies with biliary atresia, bile flow from the liver to the gallbladder is blocked. This can lead to liver damage and cirrhosis of the liver, which, if not treated, will eventually be fatal.

Newborns with this condition may appear normal at birth but jaundice develops by the 2nd or 3rd week of life. The infant may gain weight normally for the 1st month, then weight loss and irritability develop accompanied by increasing levels of jaundice. It is not known why the biliary system fails to develop normally.

Wilson's disease

Wilson's disease is an inherited disorder where there is excessive amounts of copper in the body. This causes a variety of effects, including liver disease and damage to the nervous system. Wilson's disease is a rare inherited disorder. If both parents carry an abnormal gene for Wilson's disease, there is a 25% chance that each of their children will develop the disorder (i.e., it is an autosomal recessive disease).

Wilson's disease causes the body to absorb and retain excessive amounts of copper. The copper deposits in the liver, brain, kidneys, and the eyes. The deposits of copper cause tissue damage, death of the tissues, and scarring, which causes the affected organs to stop functioning properly. Liver failure and damage to the central nervous system (brain, spinal cord) are the most predominant, and the most dangerous, effects of the disorder. If not caught and treated early, Wilson's disease is fatal.

It is most common in eastern Europeans, Sicilians, and southern Italians, but may occur in any group. The disorder most commonly appears in people under 40 years old. In children, the symptoms begin to be expressed by around 4 years old.

What is a liver transplant?

A liver transplant is the replacement of your liver with one that has been donated by someone else. The donated liver comes from someone who has died. In the future, it may be more common for donated liver tissue to come from a living person, such as a family member. In this case, you receive only a part of the donor's liver.

Liver transplantation is usually done when other medical treatment cannot keep a damaged liver functioning. About 80 to 90 percent of people survive liver transplantation. Survival rates have improved over the past several years because of drugs like cyclosporine and tacrolimus that suppress the immune system and keep it from attacking and damaging the new liver.

Who does the liver come from?

Livers come from one of two types of donor: a living relative of the person who is receiving the liver or someone who has died (a 'cadaveric donor').

A cadaveric donor is someone who has died in an accident or from a brain hemorrhage and who has been on a life support machine. Organs can only be donated once the person is 'clinically dead' which means he or she would not survive were the life support machine to be switched off. Allowing a relative's organs to be used for donation usually brings some comfort to the bereaved family.

Because the liver has the ability to regenerate, in some cases a portion of the liver from a living donor can be removed and transplanted within a recipient with liver failure. Both the donor and recipient will regenerate a liver back to normal size. This type of transplantation is increasing rapidly.

Who is suitable for a liver transplant?

A team of specially trained staff evaluates the patient to establish whether he or she would be a good candidate for a liver transplant. The staff includes people with special skills in a range of areas. The people who may be on the team include:

- Hepatologist (medical liver specialists)
- Transplant surgeons
- Social workers, psychologists, and/or psychiatrists
- Nurses
- Transplant coordinators

A liver transplant is only offered to people who have irreversible, chronic liver failure. Other medical or surgical treatments for liver problems have usually been tried before consideration of liver transplant. The liver failure may have been caused by problems such as discussed above.

Successful liver transplants have been conducted on newborn infants, children and adults including people past the age of 70.

The evaluating team considers many factors to decide whether a person should be placed on the wait list for transplantation. The person's general health and suitability for major surgery are taken into account. Risk factors are considered carefully and may result in a recommendation against transplant surgery.

Are there any risks or side effects with a transplant?

A transplant gives most people with liver failure the best quality of life possible. However, not all transplants are successful. If the transplant is successful, your child will need to take certain drugs for the rest of his or her life to prevent the liver being 'rejected'. These drugs have side effects, but there may be ways of minimizing them. Ask your doctor, nurse or pharmacist for details.

Finding the right liver

For a transplant to be successful, the tissues of the patient must be reasonably similar to those of the donor. This is called a 'good match'. In order to find out whether a kidney will be a good match or not, your child will need to have various tests to find out his or her blood and tissue type.

Blood groups and tissue types

Everyone has a blood group, which remains the same throughout life. It is based on which of two types of protein (antigens) are present in your body. The main blood groups are A, B, AB and O. People with blood group A have only antigen A in their body, people with blood group B have antigen B, people with blood group AB have both antigens, and people with blood group O have neither. The most common type is O, followed by A.

Everyone also has a tissue type, which does not change. It is based on the antigens present on the surface of most cells. You have only one tissue type but it is made up of six different parts. The three main sorts of tissue type are A, B and DR. There are about 20 different versions of the three tissue types. The more tissue types you match with the donor the less likely the body is to reject the kidney.

Preparing for transplant

Before your child joins the transplant waiting list, you will meet your transplant surgeon, hepatologist (a doctor who specializes in liver disease), and your clinical nurse specialist. They will explain all about the liver transplant and what to expect. It is also likely that you will meet the transplant social worker and a psychologist to make sure your child and your family is ready and prepared mentally for joining the list.

Waiting

If a person is a suitable candidate for a liver transplant, their name is put on a waiting list for an organ. Unfortunately, there are many more people on the wait list than there are organs available each year. There are currently more than 18,000 people in the U.S. waiting for a donor liver. Waiting time may extend several years.

People waiting for donor livers are grouped by the severity of illness and other medical factors such as blood type. Within any given group, livers are allocated based on the length of time a person has been on the wait list. A new system, the Model for End-stage Liver Disease (MELD) has recently replaced the previous 3 medical severity stages for liver transplant candidates with chronic liver diseases with a scale that goes from 6 to 40. The MELD system is based on three simple to measure laboratory tests, and the MELD score is predictive of death within 3 months (the higher the score, the higher the risk of death). Candidates with sudden, acute liver failure (status 1) are still allocated organs ahead of all other waiting patients. A system similar to MELD has been developed for children (PELD), which utilizes the same three laboratory tests in addition to a fourth blood test and a measure of growth failure.

When a suitable liver is found

People on the waiting list should be contactable at any time of the day or night. Sometimes, hospitals give out bleeps to be used only for this purpose. When you are contacted, you will need to travel to your local transplant center immediately. If traffic could be a problem, the transplant nurse could ask the police to escort you. It is important to get to the hospital as quickly as possible so that blood tests can be carried out.

The Transplant operation

If your are called into hospital, there is no guarantee that you will be able to have the transplant. There are several tests which will indicate if you are well enough for the operation. Your blood will also need to be checked to see how your body will react to the liver. If the donor is living, they will receive general anesthesia and the transplant of the liver into the recipient will occur immediately after the removal of the organ from the donor. Therefore, the patient receiving the transplant is prepared for surgery within the same time frame as the donor. They, too, will receive general anesthesia.

When a cadaveric liver becomes available, time is critical. The liver must be transplanted into the patient receiving the organ within 12 to 18 hours. A team of surgeons and anesthesiologists performs an operation to remove the liver from the donor. Additional surgical teams may be present to remove other organs. After the liver is removed from the donor, it is preserved and packed for transport. Although the donor is brain dead, this procedure is treated like any other operation using standard surgical practices and sterile techniques. Once the operation is complete and the incisions are closed, the donor's body is prepared for funeral or cremation. Organ procurement surgery respects the body and an open casket funeral is possible if desired.

In the meantime, a recipient is located and prepared for surgery. Preparation involves administration of general anesthesia. The transplant of the liver begins with an incision in the upper part of the abdomen. First, the diseased liver is removed. When the new liver is placed within the recipient, the blood vessels from the donor liver must be connected to the recipient's blood vessels. Next, the blood flow is restored. The bile duct, which carries bile made in the liver to the intestine, is also connected. After the transplant is complete, the incision is closed. The patient will begin recovery in the intensive care unit (ICU).

After Transplant Surgery

Following liver transplant surgery, the patient may remain on an artificial breathing machine for the first 24 to 48 hours of recovery. Depending on progress, some patients are moved out of the ICU in a few days. Generally, he or she will also begin eating within the week following surgery. Total hospital stay can be a little as 1 week, and is typically a couple of weeks.

Because the organ will be identified as foreign by the recipient's immune system, rejection of the new liver is always a possibility. Powerful drugs called immunosuppressants are given starting at the time of liver transplant surgery to try to prevent rejection. Within the first few weeks following transplantation, blood tests are done to confirm that correct dosage of medication is being dispensed.

Prior to discharge, the transplant team reviews information with the patient, gives instruction for follow-up care and medications, and answers the patient's questions. A prescribed rehabilitation program will continue at home including exercise, nutrition, and the continuation of immunosuppression and other medications. The signs of rejection are also discussed with the patient and family.

Living donors do not have to take any specific medications or maintain a special diet as a result of liver donation.

Returning home

At-home rehabilitation for liver transplantation is a gradual process and depends on the individual. The transplant team will give specific instructions. In general, walking is recommended to restore strength and prevent lung complications, but heavy lifting should be avoided for four to six weeks following transplant surgery. Other activities, such as driving may usually begin when the incision is free of pain. Sexual activity can resume when one is comfortable. A desire to become pregnant should be discussed ahead of time with the transplant team to determine if and when this is recommended.

Follow-up visits are required for check-ups. These begin soon after returning home. Initially, outpatient visits may occur weekly or even more often, and as time progresses the frequency of follow-up visits usually decreases.

Possible post-operative complications may arise following liver transplant surgery. They include:

- Bile duct problems
- Major bleeding
- Problems with blood vessels

Bile duct problems - Complications can arise with the connection between the donor and recipient bile duct or between the donor bile duct and intestine. If it does not heal properly, bile may leak out. Scar tissue can also block the bile duct causing bile the inability to flow.

Major bleeding - It is common for a liver transplant patient to experience bleeding after surgery. The new liver needs time to make blood-clotting proteins. Patients usually need blood transfusions, and an additional operation may be required within the first 24 to 48 hours after the transplant to resolve the problem.

Problems with blood vessels - Complications can arise with blood vessel connections between the donor liver and the recipient's blood vessels. A more serious complication is a clot in an artery or vein attached to the liver. If a clot occurs, the liver may fail.

Other problems include the long-term risks of immunosupression. These include complications related to too much or too little immunosuppression:

- Rejection
- Cancer
- Infection

Rejection - It is fairly common for a transplant patient to experience rejection episodes. The body identifies the new organ as foreign and may try to reject it. The immunosuppressive medications prevent rejection in 50 to 75% of cases. Changes may be made in the medications including an increase in dosage or the use of additional drugs to stop the rejection. Some episodes can cause permanent damage to the new liver. This may reduce longevity of the organ.

Cancer - Studies show that an estimated 6% to 8% of transplant patients will develop cancer over their lifetime with the transplant. This risk is higher than in the general population. Skin cancer is the most common, and is typically treated successfully. Some cancers result from the effects of the immunosuppressive medications and others are common cancers that occur at a higher rate in immunosuppressed individuals.

Infection - The immunosuppressant medications increase the risk of less serious and common infections such as urinary tract infection. In addition, they are associated with more serious infections like pneumonia. Finally, uncommon infections that do not affect non-immunosuppressed persons can occur. The transplant team generally follows living donors for a considerable period of time until recovery is complete.

WHAT IS PANCREAS FAILURE?

———————— ⚜ ————————

The pancreas is a five to six inch gland located behind the stomach. The pancreas produces enzymes that are used for digestion, and insulin, which is essential for life because it regulates the use of blood sugar throughout the body.

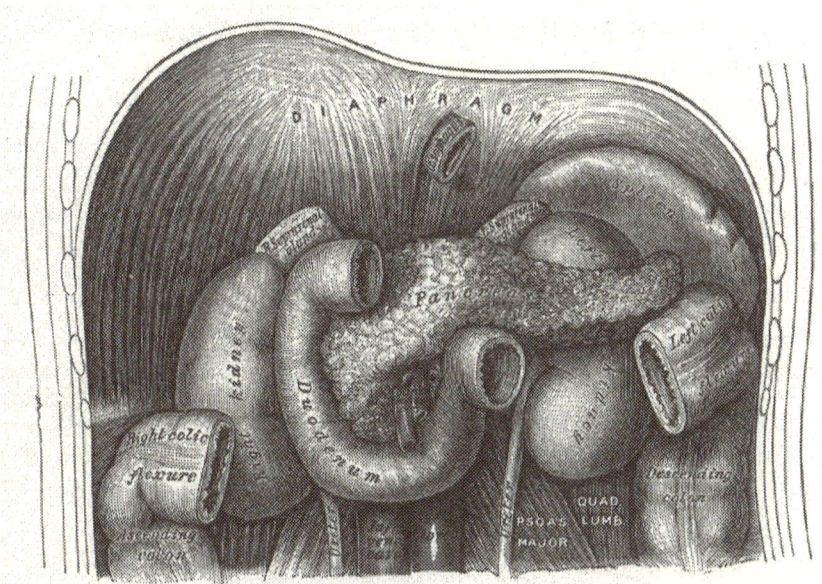

Doctors may recommend a pancreas transplant to treat diabetes. The type of diabetes that may be treated with a pancreas transplant is usually type I or juvenile onset diabetes. With this disease, the pancreas no longer produces insulin.

Severe type I diabetes is often associated with chronic renal failure. Because of this, a person may need a kidney transplant as well as a pancreas transplant. There are three types of pancreas transplant operations:

- Combined kidney-pancreas transplant

- "Pancreas after kidney" transplant, in which the pancreas is transplanted some time after a kidney has been transplanted

- Pancreas transplant alone, for patients with functioning kidneys

A pancreas transplant team evaluates the person who needs the transplant and if they are found to be suitable, his or her name is placed on the waitlist. When a donated pancreas becomes available, it is surgically removed from the donor and transplanted into the patient. The recipient's failed pancreas is not removed.

Among 326 patients who underwent pancreas transplants in 1997 and 1998 (pancreas after kidney or pancreas transplant alone), about 87% survived for at least three years afterwards. Among 1,803 patients who underwent kidney-pancreas transplants in 1997 and 1998, about 89% survived for at least three years afterwards.

Evaluating candidates for pancreas transplantation

A team of specially trained staff evaluates the patient to establish whether he or she is a good candidate for a pancreas transplant. This staff includes people with special skills in a range of areas. The people who may be on the team include:

- Nephrologists (kidney specialists) and/or endocrinologists (diabetes specialists)

- Transplant surgeons

- Social workers, psychologists, and/or psychiatrists

- Nurses

- Transplant coordinators

A pancreas transplant is only offered to people who have severe diabetes. Successful pancreas transplants have been conducted on up to the age of about 50. Pancreas transplantation in people older than 50 is less common.

The evaluating team considers many factors to decide whether a person should be placed on the waitlist for a transplant. The person's general health and suitability for major surgery are taken into account. Risk factors are also considered carefully and may result in a recommendation against transplant surgery.

A pancreas transplant would not be performed for people with certain conditions. These include:

- Most cancers, unless successfully treated at lease five years previously
- Infections that cannot be completely treated or cured, such as tuberculosis.
- Severe heart, lung, liver, or kidney problems or complications from diabetes that would make the operation too risky

It is a normal reaction of the body to reject the donated organ. Anti-rejection drugs are prescribed to prevent this rejection. The candidate must be willing to take anti-rejection medicines indefinitely to keep the body from rejecting the donor pancreas. The person will also need lifelong follow-up by healthcare professionals.

Waiting

If a person is a suitable candidate for a pancreas transplant, their name is put on a waiting list for an organ. Unfortunately, there are many more people on the wait list than there are organs available each year. There are currently more than 3,500 people in the US waiting for a pancreas or a combined pancreas-kidney transplant. Waiting time may extend several years.

People waiting for a donor pancreas are grouped by medical factors such as blood type. Within any given group, a pancreas is allocated based on the length of time a person has been on the waitlist and, especially in the case of a kidney-pancreas transplant, on degree of match between donor and recipient.

The Transplant Surgery

When a donor pancreas becomes available, time is important. The pancreas must be transplanted into the patient receiving the organ within 12 to 15 hours. A team of surgeons and anesthesiologists performs an operation to remove the pancreas from the donor. Additional surgical teams may be present to remove other organs. After the pancreas is removed from the donor, it is preserved and packed for transport. Although the donor is brain dead, this procedure is treated like any other operation using standard surgical practices and sterile techniques. Once the operation is complete and the incisions are closed, the donor's body is prepared for funeral or cremation. Organ procurement surgery respects the body and an open casket funeral is possible if desired.

In the meantime, a recipient is located and prepared for surgery as well. Preparations involve administration of general anesthesia and placement on a ventilator, or artificial breathing machine, during the surgery. An incision is made in the abdomen. In the case of a pancreas transplant alone, the pancreas can be placed on the left or the right side. In the case of a "pancreas after kidney" transplant, where the kidney transplant has already been done, the pancreas is placed on the opposite side to the kidney. In the case of a combined kidney-pancreas transplant the pancreas is often placed first, on the right side. The kidney is then placed on the left.

Blood flow needs to be restored to the new pancreas. A major artery and a major vein are connected to the new pancreas. After that, the beginning of the small intestine (called the duodenum) from the donor pancreas is connected to the recipient's intestine or bladder. After all of the connections have been completed, the incision is closed. The recipient is moved to the intensive care unit for recovery.

Partial pancreas transplantation

Unfortunately, there are not enough cadaver pancreases to go around because not enough people sign up to be organ donors, and each pancreas must meet strict guidelines. When a whole cadaver pancreas is not available, a person can receive a portion of a pancreas from a living relative.

When a patient with diabetes is receiving a kidney transplant from a living relative, it is usually beneficial to perform a partial pancreas transplant at the same time. Since the transplanted kidney will become damaged by diabetes over time, transplanting a partial pancreas from the same donor will help control blood glucose levels and protect the new kidney from further damage. Transplant success seems higher when patients and donors are matched for HLA types, and a pancreas transplanted along with a kidney is less likely to fail than a pancreas transplanted alone.

Kidney-Pancreas Transplant

Combined transplantation of the kidney and pancreas is performed for those who have kidney failure as a complication of insulin-dependent diabetes mellitus (also called Type I diabetes). Kidney and pancreas transplant candidates may be currently on dialysis or may require dialysis in the near future.

After combined transplantation of the kidney and pancreas, the kidney will be able to filter and excrete wastes so dialysis will not be needed. The transplanted pancreas will produce insulin to control the diabetes. Combined kidney and pancreas transplants and single pancreas transplants are only performed with cadaver donors.

What happens during the kidney/pancreas transplant surgery?

Kidney and pancreas transplantation involves placing a healthy kidney and pancreas into the body where they can perform all of the functions that a failing kidney and pancreas cannot.

The new kidney is placed on the lower left side of the abdomen where it is surgically connected to nearby blood vessels. Placing the kidney in this position allows it to be easily connected to blood vessels and the bladder. The vein and artery of the new kidney are attached to your vein and artery. The new kidney's ureter is attached to your bladder to allow urine to pass out of your body.

The new pancreas is placed on the lower right side of your abdomen where it is surgically connected to nearby blood vessels. The vein and artery of the new pancreas are attached to your vein and artery.

What is the success rate of the double transplant?

After the double transplant is performed, there is an 80 to 85 percent chance that the patient will require no insulin and no dialysis for one year. In addition, there is a 70 percent chance that this success will continue over the next 5 years.

What are the benefits of double organ transplantation?

A successful kidney and pancreas transplant gives you increased strength, stamina and energy. After transplantation, you should be able to return to a more normal lifestyle and have more control over your daily living. You can have a normal diet and more normal fluid intake.

If you were dependent on dialysis before the transplant, you'll have more freedom because you won't be bound to your dialysis schedules. The pancreas transplant will keep your blood sugar normal. Frequently after transplantation, your blood sugar level before eating will be 90 or less; after a meal it may reach 140 -- all without insulin. Further complications of diabetes may be delayed with better blood sugar control.

What are the risks of double organ transplantation?

Since two organs are transplanted, the risk of surgical complications is about twice that of a single organ transplant (such as a kidney-only transplant). Since the pancreas is joined to the bladder during the operation, some loss of fluids occurs. You may need to drink more than usual after the transplant surgery in order to prevent dehydration.

There is also a risk of rejection after any type of transplant surgery. Rejection is your body's way of not accepting the new kidney and pancreas. Since your body recognizes the new organs as foreign objects, it will normally try to get rid of them or "reject" them.

However, you are given medications to prevent rejection. You will need to take these medications for life and have your blood work drawn as scheduled to prevent rejection episodes.

After Transplant Surgery

Following the pancreas transplant surgery, the recipient is cared for in the hospital for 7 to 10 days. It is common for the patient to be able to get out of bed and start walking within 24 to 48 hours of the transplant.

Because the organ will be identified as foreign by the recipient's immune system, rejection of the new pancreas is always a possibility. Powerful drugs called immunosuppressants are given starting at the time of pancreas transplant surgery to try to prevent rejection. Within the first few weeks following transplantation, blood tests are done to confirm that correct dosage of medication is being dispensed.

Prior to discharge, the transplant team reviews information with the patient, gives instructions for follow-up care and medications, and answers the patient's questions. A prescribed rehabilitation program will continue at home including exercise, nutrition, and the continuation of immunosuppression and other medications. The signs of rejection are also discussed with the patient and family.

Returning Home

At-home rehabilitation is a gradual process, and depends on the individual. The transplant team will give specific instructions. In general, walking is recommended to restore strength, but heavy lifting and straining should be avoided for several weeks following surgery. Other activities, such as driving may usually begin when the incision heals. Sexual activity can resume when one is comfortable. A desire to become pregnant should be discussed ahead of time with the transplant team to determine if and when this is recommended.

Follow-up visits are required for check-ups. These begin soon after returning home. Initially, outpatient visits may occur weekly or even more often, and as time progresses the frequency of follow-up visits usually decreases.

Possible post-operative complications may arise following pancreas transplant surgery. They include:

- Clotting of Major Vessels
- Major Bleeding
- Leaking from intestinal connection
- Pancreatitis

Clotting of major vessels - Blood is supplied to the pancreas by a major artery and vein. The artery or vein may become blocked, or clots may form. This can cause sudden pancreas failure. In that case the new pancreas will need to be removed.

Major bleeding - While major bleeding is not common after pancreas transplantation, there are some cases where small blood vessels bleed. These blood vessels in the donor pancreas need to be tied off during surgery. In some cases a second operation is performed to control bleeding and remove any blood clots.

Leaking from intestinal connection - The bowel must heal together after surgery. If it doesn't, leakage and infection may occur. This usually requires another operation for repair and often the pancreas has to be removed as a result.

Pancreatitis - Inflammation of the donor pancreas may cause fluid accumulation in the abdomen, pain or abnormal function of the donor pancreas.

Other problems include the long-term risks of immunosupression. These include complications related to too much or too little immunosuppression:

- Rejection
- Cancer
- Infection

Rejection - It is fairly common for a transplant patient to experience rejection episodes. The body identifies the new organ as foreign and may try to reject it. The immunosuppressive medications prevent rejection in 50 to 75% of cases. Changes may be made in the medications including an increase in dosage or the use of additional drugs to stop the rejection. Some episodes can cause permanent damage to the new pancreas. This may reduce longevity of the organ.

Cancer - Studies show that an estimated 6% to 8% of transplant patients will develop cancer over their lifetime with the transplant. This risk is higher than in the general population. Skin cancer is the most common, and is typically treated successfully. Some cancers result from the effects of the immunosuppressive medications and others are common cancers that occur at a higher rate in immunosuppressed individuals.

Infection - The immunosuppressant medications increase the risk of less serious and common infections such as urinary tract infection. In addition, they are associated with more serious infections like pneumonia. Finally, uncommon infections that do not affect non-immunosuppressed persons can occur.

WHAT IS LUNG (PULMONARY) FAILURE?

————— ⚜ —————

The lungs are a pair of highly elastic and spongy organs in the chest. They are the main organs involved in breathing. They take in air from the atmosphere and provide a place for oxygen to enter the blood and for carbon dioxide to leave the blood. The lungs are divided into sections, with three on the right and two on the left.

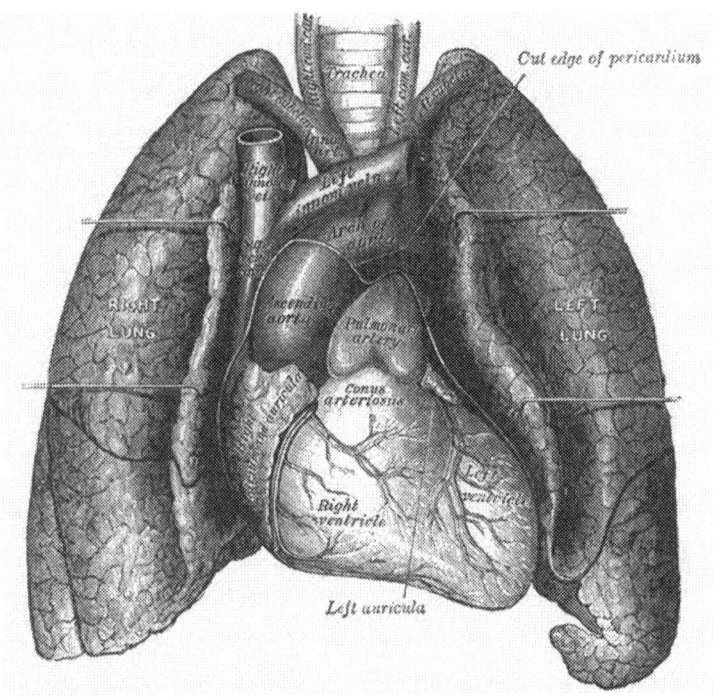

Evaluating candidates for lung transplantation

A team of specially trained staff evaluates the patient to establish whether he or she would be a good candidate for a lung transplant. The staff includes people with special skills in a range of areas. The people who may be on the team include:

- Pulmonologists (medical lung specialists)
- Transplant surgeons
- Social workers, psychologists, and/or psychiatrists
- Nurses
- Transplant coordinators

A lung transplant is only offered to people who have irreversible lung failure. These people will not live longer than 1 to 2 years unless they receive a lung transplant. Other medical or surgical treatments for cardiac problems have usually been tried before consideration of lung transplants. The lung failure may have been caused by problems such as:

- Emphysema, a chronic lung disease
- Pulmonary fibrosis, a group of respiratory diseases that are associated with scarring
- Cystic fibrosis, an inherited disease that affects the respiratory and digestive systems
- Alpha-1-antitrypsin deficiency, a deficiency of a protein produced in the liver that is associated with emphysema and liver disease
- Pulmonary hypertension, a condition in which pressure in the blood vessels of the lungs increases. This increased pressure causes damage to the blood vessels and heart.

Emphysema

Emphysema is a condition in which there is over-inflation of structures in the lungs known as alveoli or air sacs. This over-inflation results from a breakdown of the walls of the alveoli, which causes a decrease in respiratory function (the way the lungs work) and often, breathlessness. Early symptoms of emphysema include shortness of breath and cough. Emphysema and chronic bronchitis together comprise chronic obstructive pulmonary disease (COPD).

Pulmonary fibrosis

Pulmonary Fibrosis involves scarring of the lung. Gradually, the air sacs of the lungs become replaced by fibrotic tissue. When the scar forms, the tissue becomes thicker causing an irreversible loss of the tissue's ability to transfer oxygen into the bloodstream.

Cystic fibrosis

Cystic fibrosis (CF) — a life-threatening disorder that causes severe lung damage and nutritional deficiencies — used to be a genetic mystery, and most people with the disease didn't live beyond their teens. But researchers have made progress in unraveling the genetic basis of CF, which has led to earlier detection. In addition, improved and more consistent treatments now allow people with CF to live into their 30s and 40s and even beyond, and to have fuller, more comfortable lives.

CF is an inherited (genetic) condition affecting the cells that produce mucus, sweat, saliva and digestive juices. Normally, these secretions are thin and slippery, but in CF, a defective gene causes the secretions to become thick and sticky. Instead of acting as a lubricant, the secretions plug up tubes, ducts and passageways, especially in the pancreas and lungs. Respiratory failure is the most dangerous consequence of CF.

Alpha-1-antitrypsin deficiency

Alpha-1-antitrypsin protects the delicate tissues of the lungs by binding to neutrophil elastase, a protein released by white blood cells that digests bacteria and other foreign objects in the lungs. When a person who is deficient of Alpha-1-antitrypsin inhales irritants, or contracts a lung infection, the neutrophil elastase released in the lungs continues digesting irritants unchecked, eventually destroying healthy lung tissue. The eventual result of the destruction of healthy lung tissue by neutrophil elastase is emphysema.

However, alpha-1-antitrypsin deficiency emphysema (also known as "Genetic" or "Inherited" Emphysema) is different than emphysema caused by smoking ("acquired" emphysema). In emphysema caused by smoking the damage usually affects the upper portion of the lungs. In the Alpha-1 patient, the lower regions of the lungs are first affected. With either cause, the lungs are hyper inflated due to air trapping caused by the destruction of the lung tissue, and the diaphragms are flattened due to the hyperinflation of the lungs.

Many people with Alpha-1 also have chronic bronchitis. With this, the lung lining becomes swollen and congested with mucus, restricting airflow. The bronchi (air passages) often go into bronchospasms, which are contractions of the muscles, which further reduce airflow. This often results in a chronic cough.

Pulmonary hypertension

Pulmonary hypertension is a rare blood vessel disorder of the lung in which the pressure in the pulmonary artery (the blood vessel that leads from the heart to the lungs) rises above normal levels and may become life threatening.

Symptoms of pulmonary hypertension include shortness of breath with minimal exertion, fatigue, chest pain, dizzy spells and fainting. When pulmonary hypertension occurs in the absence of a known cause, it is referred to as primary pulmonary hypertension (PPH). This term should not be construed to mean that because it has a single name it is a single disease. There are likely many unknown causes of PPH. PPH is extremely rare, occurring in about two persons per million population per year.

Successful lung transplants have been conducted on newborn infants, children and adults including people past the age of 60.

The evaluating team considers many factors to decide whether a person should be placed on the wait list for a transplant. The person's general health and suitability for major surgery are taken into account. Risk factors are considered carefully and may result in a recommendation against transplant surgery.

A lung transplant would not be performed for people with certain conditions. These include:

- Most cancers, unless successfully treated at least five years previously
- Infections that cannot be completely treated or cured, such as tuberculosis
- Severe heart, liver or kidney problems that would make the operation too risky

It is a normal reaction of the body to reject the donated organ. Anti-rejection drugs are prescribed to prevent this rejection. The candidate must be willing to take anti-rejection medicines indefinitely to keep the body from rejecting the donor lung. The person will also need lifelong follow-up by health care professionals.

Heart-Lung Transplant

Heart-lung transplant may be recommended for patients with:

- Severely diseased lungs, such as primary pulmonary hypertension
- Severely damaged heart

Heart-and-lung transplant operations have been performed since 1980 in the United States. In 1997, approximately 150 of these operations were performed. The donated heart and lungs are from a person who has been declared brain-dead, but remains on life-support. Tissue matches must be performed to assure the patient's best chance of not rejecting the transplanted organs.

While the patient is deep asleep and pain-free (general anesthesia), an incision is made through the breast bone (sternum). Tubes are used to re-route the blood to a heart-lung bypass machine. This keeps the blood oxygenated and circulating during the surgery.

The patient's heart and lungs are removed, and the donor heart and lungs are stitched into place.

Waiting

If a person is a suitable candidate for a lung transplant, their name is put on a waiting list for an organ. Unfortunately, there are many more people on the wait list than there are organs available each year. There are currently more than 3,800 people in the U.S. waiting for a donor lung. Waiting time may extend several years.

Lungs are allocated based on the length of time a person has been on the waitlist and blood type.

The Transplant Surgery

When a donor lung becomes available, time is critical. The lung must be transplanted into the patient receiving the organ within 4 to 6 hours. A team of surgeons and anesthesiologists performs an operation to remove the lung from the donor. Additional surgical teams may be present to remove other organs. After the lung is removed from the donor, it is preserved and packed for transport. Although the donor is brain dead, this procedure is treated like any other operation using standard surgical practices and sterile techniques. Once the operation is complete and the incisions are closed, the donor's body is prepared for funeral or cremation. Organ procurement surgery respects the body and an open casket funeral is possible if desired.

Typically, both lungs are removed from the donor together. If the recipient is in need of a double lung transplant, both lungs will be transplanted. Otherwise, the lungs are usually separated after they are removed from the donor and used for two single lung transplant recipients.

In the meantime, a recipient is located and prepared for surgery as well. Preparation involves administration of general anesthesia, and placement on an artificial breathing machine. The transplant of the lung begins with removal of the diseased lung and the blood vessel attachments to the heart and large airway (bronchus). When the lung is placed within the recipient, the blood vessels and bronchus from the donor lung must now be connected to the recipient's corresponding blood vessels and bronchus. Next, the blood flow and airflow are restored. After the transplant is complete the incision is closed. The patient will begin recovery in the intensive care unit (ICU).

When a double lung transplant is performed, it is much like two single lung transplants. The lung that is more diseased is transplanted first and then the less diseased lung is transplanted.

After Transplant Surgery

Following lung transplant surgery, the patient may remain on an artificial breathing machine for the first 12 hours of recovery. However, if the donor lung is functioning properly, the artificial breathing machine may be removed at the end of surgery. Depending on progress, some patients are moved out of the ICU in a few days. Generally, he or she will also begin eating within the week following surgery. Total hospital stay is typically 7 to 10 days.

Because the organ will be identified as foreign by the recipient's immune system, rejection of the new lung is always a possibility. Powerful drugs called immunosuppressants are given starting at the time of lung transplant surgery to try to prevent rejection. Within the first few weeks following transplantation, blood tests are done to confirm that correct dosage of medication is being dispensed.

Prior to discharge, the transplant team reviews information with the patient, gives instructions for follow-up care and medications, and answers the patient's questions. A prescribed rehabilitation program will continue at home including physical activity, breathing exercises, nutrition, and the continuation of immunosuppression and other medications. The signs of rejection are also discussed with the patient and family.

Returning Home

At-home rehabilitation for lung transplantation is a gradual process, and depends on the individual. The transplant team will give specific instructions. In general, walking is recommended to restore strength and prevent lung complications, but heavy lifting should be avoided for four to six weeks following transplant surgery. Other activities, such as driving may usually begin when the incision heals. Sexual activity can resume when one is comfortable. A desire to become pregnant should be discussed ahead of time with the transplant team to determine if and when this is recommended.

Follow-up visits are required for check-ups. These begin soon after returning home. Initially, outpatient visits may occur weekly or even more often, and as time progresses the frequency of follow-up visits usually decreases.

Possible post-operative complications may arise following lung transplant surgery. They include:

- Major bleeding
- Pneumonia
- Pulmonary edema

Major bleeding - Rarely the patient may experience bleeding after surgery. Patients may require an additional operation to resolve the bleeding problem or remove blood clots.

Pneumonia - Infection in the transplanted lung can be life-threatening.

Pulmonary edema - The donor lung may accumulate fluid that prevents the effective exchange of oxygen and carbon dioxide.

Other problems include the long-term risks of immunosupression. These include complications related to too much or too little immunosuppression:

- Rejection
- Cancer
- Infection

Rejection - It is fairly common for a transplant patient to experience rejection episodes. The body identifies the new organ as foreign and may try to reject it. The immunosuppressive medications prevent rejection in 50 to 75% of cases. Changes may be made in the medications including an increase in dosage or the use of additional drugs to stop the rejection. Some episodes can cause permanent damage to the new lung. This may reduce longevity of the organ.

Cancer - Studies show that an estimated 6% to 8% of transplant patients will develop cancer over their lifetime with the transplant. This risk is higher than in the general population. Skin cancer is the most common, and is typically treated successfully. Some cancers result from the effects of the immunosuppressive medications and others are common cancers that occur at a higher rate in immunosuppressed individuals.

Infection - The immunosuppressant medications increase the risk of less serious and common infections such as urinary tract infection. In addition, they are associated with more serious and potentially life-threatening infections like pneumonia. Finally, uncommon infections that do not affect non-immunosuppressed persons can occur.

Every transplant, no matter what organ—heart, lung, liver, or kidney—is a partnership between the transplant center and you to keep you and your new transplant healthy.

LIFE AFTER TRANSPLANTATION

———————— ⚜ ————————

After receiving your transplant, you will continue to work closely with your transplant team, who will play an active role in your recovery. Although living with a transplant will give you a new lease on life, caring for a healthy organ involves taking sensible steps to recover fully and return to a normal, active lifestyle.

Preventing Rejection

Once you have received your organ, you should need to do everything possible to stay healthy and prevent rejection. While the risk of rejecting your new organ decreases as time goes on, it never goes away.

What Is Rejection?

Rejection is the process by which the organ recipient's immune system recognizes, becomes sensitized against and attempts to eliminate the foreign antigens of the donor organ. It often occurs when your immune system detects things like bacteria or a virus. Some degree of rejection occurs with every transplant, but how clinically significant the rejection is very individual.

Although acute rejection, the body's attempt to destroy the transplanted organ, is possible years after a transplant, at least one episode is common within the first year after a transplant. Despite the use of immunosuppression therapy, acute rejection can occur and often lead chronic rejection. Chronic rejection, which is characterized by gradual loss of organ function, is an ongoing concern for transplant recipients because it can occur weeks, months or years after transplantation. Therefore, organ recipients should be aware of the signs of both acute and chronic rejection. Call your doctor as soon as you experience any of them. Symptoms include:

- Pain or tenderness over the transplant site
- Fever
- Flu-like symptoms such as chills, nausea, vomiting, diarrhea, tiredness, headache, dizziness and body aches and pains
- Change in pulse rate
- Weight gain
- Swelling
- Less urine

Anti-Rejection Medication

The goal of anti-rejection medicines, also called immunosuppressants, is to adequately suppress the immune response to prevent rejection of the transplanted organ while maintaining sufficient immunity to prevent overwhelming infection. Many of the medications used to achieve immunosuppression have adverse effects of their own. That's why a combination of medications that work in different phases of the immune response minimize side effects and produce effective immunosuppression. Clinical immunosuppression usually occurs in three phases: induction, maintenance and anti-rejection.

Medications

After transplants, the focus for patients transitions from waiting for a donor organ to learning how to manage medications and their side effects as part of daily life. One of the most important aspects of protecting your transplant is the medications prescribed to you.

The most critical part of transplantation is preventing rejection of the grafted organ.

- Different transplant centers use different drug combinations to fight rejection of a transplanted kidney.

- The drugs work by suppressing your immune system, which is programmed to reject anything "foreign," such as a new organ.

- Like any medication, these drugs can have unpleasant side effects.

- Some of the most common immune-suppressing drugs used in transplantation are described here.

    - Cyclosporine: This drug interferes with communication between the T cells of the immune system. It is started immediately after the transplant to suppress your immune system and continued indefinitely. Common side effects include tremor, high blood pressure, and kidney damage. These side effects are usually related to the dose and can often be reversed with proper dosing.

    - Corticosteroids: These drugs block T-cell communication as well. They are usually given at high doses for a short period immediately after the transplant and again if rejection is suspected. Corticosteroids have many different side effects, including easy bruising of the skin, osteoporosis, avascular necrosis (bone death), high blood pressure, high blood sugar, stomach ulcers, weight gain, acne, mood swings, and a round face. Because of these side effects, many transplant centers are trying to reduce the maintenance dose of the drug as much as possible or even to replace it with other drugs.

- Azathioprine: This drug slows the production of T cells in the immune system. It usually is used for long-term maintenance of immunosuppression. The most common side effects of this drug are suppression of the bone marrow, which produces blood cells, and liver damage. Many transplant centers are now using a newer drug called mycophenolate mofetil instead of azathioprine.

- Newer antirejection drugs include tacrolimus, sirolimus, and mizoribin, among others. These drugs are now being used to try to reduce side effects and to replace drugs after episodes of rejection.

- Other costly and experimental treatments include using antibodies to attack specific parts of the immune system to decrease its response.

Diet and Exercise

Transplant recipients need to be aware of the important role of a healthy diet and exercise plan in healing. Prior to your discharge from the hospital, talk to your doctor or dietitian about your goals and requirements. Since each person is different, know that you can ask for help in developing a plan that fits your needs, likes and dislikes.

Diet After a Transplant

After your transplant, you will be feeling better and looking forward to returning to your normal lifestyle. A major part of that normal lifestyle is the ability to enjoy eating again. However, some of the drugs you will be taking after your transplant affect the way your body processes food. This may cause you to eat more, causing you to gain weight.

Excessive weight gain can be harmful to your health because it increases the risk of heart disease, diabetes and high blood pressure. The following tips may help you avoid unwanted weight gain:

- Eat a balanced diet with plenty of fruits and vegetables.
- Eat a minimum amount of salt, processed foods and snacks.
- Use herbs and spices to add flavor, instead of salt.
- Watch your food intake and drink plenty of water (unless you are told to limit fluids).
- Try to eat high-fiber foods, such as raw vegetables and fruits, which make you feel full.
- Add calcium to your diet by eating calcium-rich foods, such as low-fat dairy products and green, leafy vegetables or calcium supplements.
- Eat as little fat and oil as possible.
- Read food labels so that you can be smart when food shopping.
- Become more aware of serving sizes that are listed on food labels.
- Because protein helps your build muscles and tissue, which will help you heal after surgery, eat foods high in protein, such as meat, poultry (i.e. chicken), fish, eggs, nuts (without salt) and beans.
- Select healthier condiments, such as mustard, and low-fat mayonnaise and salad dressings.
- Choose healthy cooking methods. Instead of frying, try baking, grilling, broiling or steaming foods. And instead of oil, use nonstick, fat-free spray or sauces.
- When dining out, try to eat smaller portions and avoid high-fat entrees.

- Don't drink alcohol or use any drugs that aren't prescribed by your physician, as these may harm your new organ. If you have a problem with drugs or alcohol, talk with your social worker, who can help arrange for counseling and other support services.

Exercise After a Transplant

Most people are weak after any surgery. Transplant recipients must recover from surgery, as well as the illness that caused the need for a transplant. As a result, exercise and muscle strain should be limited when you return home. Talk with your doctor about what to expect.

As you start to feel better, regular exercise will help you regain your strength. Because you may feel tired at first, you should take rest breaks during exercise. Gradually, increase the amount and type of physical activity you enjoy.

Health Concerns

What happens after transplantation depends on the organ transplanted and the recipient's specific medical situation. Most patients recover fully, return to work and resume a normal, active life after receiving a new organ. However, there is a possibility of developing unrelated health problems after transplantation. That's why it is important to work closely with your doctor concerning your overall wellness, as well as regarding the following health concerns:

Anxiety and Depression

Patients and their families face a new lifestyle after transplantation that may cause them to feel nervous, stressed or depressed. Because emotional and psychological support is a continuing process, ask your social worker about counseling services that can help you and your family deal with these changes. Professionals can help you work through concerns about your self image; mood swings;

job planning; rehabilitation; family stresses, such as parent-child conflicts, marital conflict or changes in sexual functioning; and financial concerns, such as questions about Medicare, disability or insurance.

Diabetes

Some anti-rejection medicines are known to cause high blood sugar. Although it is typically a temporary condition after transplantation, it is more common in patients who have a family history of diabetes and patients who are over weight. It can be controlled by reducing the dose of a patient's anti-rejection medicines or changing medications all together.

GI Upset

GI (gastrointestinal) upset is also a common complaint after a transplant. Patients on steroid therapy may be at an increased risk of developing ulcers due to increased hydrochloric acid from the stress of the procedure. Treatment of GI Upset may include one or a combination of drugs that reduce acid production. In addition, people with GI upset should take several steps to reduce symptoms, including:

- Reducing the intake of caffeine, alcohol and other over-the-counter medications that cause GI upset.
- Taking medications with food to decrease irritation.

Gout

Gout is a painful and potentially disabling form of arthritis. Diagnosing gout can be difficult and treatment plans vary based on a patient's existing medical problems and medications.

High Cholesterol

Many immunosuppressant drugs can contribute to high cholesterol. This condition therefore affects many transplant recipients. When a patient develops high cholesterol, blood vessels, including the ones attached to the transplanted organ, become clogged, which affects the flow of blood. This slowing of blood flow can affect the success of your transplant and may even lead to heart disease. It is important to talk to your doctor about how to reduce the risk factors of heart disease, including controlling your cholesterol.

Hypertension

Hypertension, or high blood pressure, is common immediately after transplant. Certain anti-rejection medications, as well as the original disease, all can contribute to hypertension. Treatment of hypertension may include one or a combination of drugs, and often, as anti-rejection medicines are tapered to a maintenance dose, hypertension may decrease. Talk to your doctor about what's right for you and how to avoid high blood pressure.

Sexual Relations

Sexual concerns after transplantation are commonly experienced, yet seldom discussed or addressed during evaluation. It is therefore very important to talk with your doctor about your sexual history and concerns.

However, sexual function and interest can be related to how well your body has accepted your new organ and how realistic your expectations were for life after your transplant. A counselor can also help a couple understand the difference between pre- and post-transplant problems. In addition, support groups may lend the emotional support surrounding any changes in sexual relations.

Additionally, in sexual relations, as in all issues, recipients must remember that they are immune suppressed and subject to many kinds of infections. In fact, some infections in recently transplanted patients can be potentially life threatening. Consequently, it is important to consider the following points:

- The sharing of saliva during kissing can expose both partners to active diseases, such as colds or other viruses.

- Condoms don't prevent diseases that are spread by contact between the area surrounding the penis and external genitals.
- The risk of contracting infectious diseases though oral sex is possible, especially if ejaculation occurs or if there are any sores or wounds on either partner.

Dental care after a transplant

Kidney transplant patients usually have few dental problems as long as their new kidney is working properly, although your child will need to see the dentist every six months for a check-up. There are a couple of problems which tend to occur in people who have had kidney transplants, which are due more to the anti-rejection drugs than the transplant itself.

- **Gum overgrowth (gingival hypertrophy):** This is a common problem in people taking cyclosporin. Perfect dental hygiene will reduce the overgrowth but does not prevent it. If your child's gums overgrow, it can be treated with intensive cleaning, use of a special spray or gel twice a day and a small operation.
- **Infections:** These can be serious because a simple infected tooth or gum can spread to the whole body. It is important to prevent this from happening by recognizing the problem early and getting prompt treatment from your dentist.

Instructions for perfect dental hygiene:

- Your should brush and floss teeth twice a day. You will be shown how to do this properly.
- You should see the dentist every three to six months for a check-up and to prevent problems.
- If you develop pain or a sore area in their mouth, you should visit the dentist as soon as possible.

STORIES OF SUCCESS

————————— ✤ —————————

Mary Lou Drittler
Lung Transplant Recipient, 2002
"Back to Her Normal Life"

Mary Lou Drittler had suffered the effects of emphysema since 1991, when she was diagnosed with the disease. By 1996 she was on oxygen most of the time and was able to do very little. She lost her breath after walking just a few steps, and led a mostly sedentary life. When shopping and running errands, she traveled in a wheelchair.

After her pulmonologist told her she could benefit from a lung transplant, Drittler explored her options. She considered several hospitals and even looked into Lung Volume Reduction as an option, for which she was not a candidate.

Finally, in December 2001 she seriously looked into lung transplantation. One area hospital turned her down right away. "I very quickly got a letter back telling me I was too old; they wouldn't even talk to me. They have a very strict limit that you have to be 63 to apply and they did not operate after 65. I was 66 at the time," recalled Drittler, who resides in Silver Spring, Md.

After that, she decided to go to the University of Pittsburgh, because Bartley Griffith, M.D., director of Cardiac Surgery and Cardiothoracic Transplantation at the University of Maryland Medical Center (UMMC), had performed a lung transplant on a friend of hers there nine years earlier.

When she learned that Dr. Griffith was coming to the UMMC, she decided to follow him here and get her transplant at the Maryland Heart Center. She met with Dr. Griffith and Jo Ann Sikora, a cardiac surgery nurse practitioner, and then met with the whole transplant team in January 2002. "They determined that I was a good candidate because I was in generally good health other than the lung," said Drittler.

She officially went on the transplant waiting list on April 29, 2002, but she didn't have to wait long -- two months later she had a lung donor. Dr. Griffith performed the lung transplant on June 27, 2002. He said Drittler's operation was a success.

"Her overall wellness speaks for itself. She took on a high-risk operation and we took on a high-risk patient and together we won," said Dr. Griffith, who is also professor and head of the Division of Cardiac Surgery at the University of Maryland School of Medicine. "I think it speaks to a number of things -- her heartiness and the fact that this program is full of physicians, nurses and caregivers who will go the extra mile."

Since the transplant, Drittler's condition has improved dramatically. She is participating in an outpatient pulmonary rehabilitation program, and she exercises regularly at home on her treadmill.

"Now I feel good and I'm back to my normal life," said Drittler. "I'm a lot more active. I spend time with the grandchildren, go shopping, and I join all these transplant groups so I can go to a lot of parties."

In fact, Drittler recently attended an annual holiday party held for UMMC heart and lung transplant recipients, their families and the doctors and nurses who took care of them.

She appreciates the care she received at the Maryland Heart Center. "I couldn't have gone anywhere where I could have gotten better treatment," Dritter said. "The whole transplant team is really caring and on top of things."

And Drittler is no stranger to UMMC. At one time, one of Drittler's daughters was also a patient at the Medical Center. "My older daughter had breast cancer and she had a stem cell transplant done here and at that time I felt the same way. The care is unbelievable," she said.

Since the transplant, family members have noticed a change in Drittler.

"She's doing so much more," said Michele Burhofer, one of Drittler's daughters. "Before (the transplant), she wouldn't come over to my house at all. Now she comes over and spends a weekend and just enjoys herself. She has a new appreciation for life."

Her caregivers have also noticed a difference. As Sharon Lesser, a pulmonary clinical nurse at the Medical Center observed, "Mary Lou looks wonderful and her spirits are great. Every time I see her she has a bigger and brighter smile on her face."

Steve Autrey
Liver Recipient, 1998
In Sickness and In Health... A Young Couples Love Story.

The year was 1995; this love story begins on a clear autumn night in San Antonio, Texas. A young man named Steve Autrey saw a tall young college student leaning up against the bar in a local country-western club. When his glaze caught hers he knew then she was like no girl he had met up to that point. "They were made for each other" Steve's mother likes to say. The two fell in love and where married a year later. They graduated from college and moved to Austin, TX, where Steve got a job as an accountant.

Like any young couple, they had plans of building a home and starting a family, but Fate had other plans. Steve was a 25 year old newly married young man who found himself in need of a liver transplant. Steve had developed cirrhosis of the liver as a complication of underlying Crohn's disease.

Steve's story starts as any young boys. His was a life full of dreams and aspirations. As a boy he seemed to be susceptible the stomach flu. At age 15, he was diagnosed with Crohn's disease, a condition of chronic inflammation of the intestines. There is no known cure for Crohn's disease, but hope is to control it with medications. Steve, like myself, never gave much thought to being sick. His father remembers him getting up early while in high school to be a swim practice at dawn everyday. Steve suffered from a chronic anemia (low blood count) because of his Crohn's disease, which caused him to not have the stamina of other "healthy" teenagers. However, Steve didn't let this stop him.

In 1991, Steve started at Texas A&M University. As all freshmen's do, Steve ate poorly, didn't sleep much, and went to frat parties. Although trying to do all the things a young freshman in college is expected to do, his Crohn's would not allow him to recover as a "healthy" freshman would.

By May, he was sick, weak, and bedridden. He was 5 foot 11 1/2 inches, but weighed only 95 pounds. His father came to his dorm where he carried Steve from his bed to the hospital. The bad news was just beginning. The doctors told Steve and his parents that he had extensive liver damage and would need a liver transplant in the next five years.

Steve went back to finish college where he met Alicia, the women of his dreams. Steve and Alicia had been married just over a year when the other shoe-dropped. Steve came down with another "stomach-bug", or so he thought. He began running a high fever and coughing up blood. Steve wound up in the hospital, again. The five years had come and gone, Steve needed his liver transplant. The time was now. Without a new liver, Steve would die.

Alicia was devastated. She, as most of us, found herself scared and furious. Why her, why Steve, why now. As most transplant patients who find themselves facing a transplant, the families are equally affected. The "whys" are normal. Steve coped in his own way. He tried not to talk about his symptoms: night sweats, bleeding easily, painful swelling in his legs and abdomen, the confusion at time from the toxins building in his blood stream. Alicia kept up with all the information she could, even joking about going to medical school when the whole ordeal passed, because of the education she was getting. The folder that had held her weeding plans just over a year earlier now held the information it took to help keep her husband from dying.

Like all patients' on the waiting list, fund raising, becomes part of your fight for life. Transplant surgery is expensive, and although there is help, it does not usually cover the complete cost. Texas' U.S. Senators Phil Graham and Kay Bailey Hutchison, and former Governor Ann Richards donated autographed memorabilia to be auctioned off the help Steve.

All this time, Steve and Alicia had to hold to faith that everything would be all right. They still believed in happy endings, that "fate would not have brought them together to separate them so soon."

Eight months passed, Steve's day finally came. He received that infamous phone call, the phone message that all patients on the waiting list are living for. At 3am they started to Hermann Hospital in Houston, Texas. They had never had car trouble before until that morning. As if fate had played to cruelest of tricks they where stranded on the side of the road. Every second, every minute that passed seemed like hours as Steve's second chance at life seemed to waste away. Just as panic was starting to set in, the gleam of headlights came in the distance. The driver, a Blue Bell Ice Cream worker was on his way to work. Steve flagged him down; he drove him to a gas station where they called the sheriff who arranged to transport Steve via ambulance to the hospital.

As they wheel Steve into the operating room there is barley time for rushed "I love you's". Alicia had promised to be strong, but the scene overwhelmed her. She began to cry. There is nothing to do now but wait.

As she waited in the critical care waiting with the rest of the family, the surgeons operated skillful to take out Steve's disease ridden liver and place a new healthy liver in his body. After many hours Steve is out of surgery later that night. Alicia goes in to see him in the intensive care unit and runs her fingers through his messy hair, "even with all this mess, he's still handsome," she softly states. The tears come fresh and new, but now they are more tears of joy then apprehension and sadness, although she knows the next 24 hours will be critical.

By noon the next day, Steve is off the respirators. By one in the afternoon, Steve is able to speak with his family for the first time since going into surgery. Steve spent time recovering and is still practicing as an accountant. At the time of this interview, 2004, he is still doing well and thankful for everyday. He and Alicia are living their dreams; they have a home and family as planned. They are thankful for the angels in their lives—the Blue Bell worker, the sheriff deputy, the ambulance service, the surgeons, and most of all for the anonymous donor and the donor's family for turning their tragedy into a new chance for Steve to live his dreams.

Mitchell Steppe
Double Lung Recipient, 2003

Mitchell Steppe journey started 15 years ago when he first noticed himself getting short of breathe. Doctors told him he had emphysema, but he passed on disability and kept working at his textile job. Mr. Steppe was a man who loved to hunt, fish, and swim. But he noticed himself getting more short of breathe with activity. Mr. Steppe likes to compare his activity level to that of the Tasmanian devil, "I would put him to shame," Mr. Steppe stated in a 2002 newspaper article.

His breathing capacity slowly diminished over time until simple tasks like, showering and getting dressed became a chore. Finally, in the late 90's doctors told him he had the lungs of a 95-year-old man. The initial idea of a lung transplant was placed on hold due to the fact they thought he lung cancer. This was ruled out and he was reconsidered for a transplant.

When I spoke with him earlier last year he recounted how many trips in and out of the hospital he had during his wait. He jokes how he knew when the ambulance needed its tires rotated before the medics did because he rode it so frequently. He wife joked that they visited so many different hospital, because no matter where they were he would wind up in the hospital. She would tease him about going to a new town and ride their rescue truck for the weekend.

While waiting on the list Mr. Steppe would attend the cardio-pulmonary rehab at the local hospital to build up his endurance. During this time he was on home oxygen. Chronic Obstructive Pulmonary Disease or COPD is exactly what it sounds it is a chronic deterioration of the lung. When lung function decline to 10% capacity you die, Mr. Steppe, when he was placed on the transplant list was at 15%.

Mr. Steppe feels this last ten years might actually have made him a better person. "I have learned how to deal with my life." He explains the feeling of end stage COPD as, "it feels like somebody's trying to smother you with a pillow." Mr. Steppe deteriorated to the point where he could take a shower without using his oxygen tank to help get is breathe between washing. He lost weight and became frail.

Finally, he got the awaited telephone call. The doctors at Duke University Medical Center performed a double lung transplant on July 8th, 2003. The operation lasted 8 hours. He felt at peace undergoing his transplant, "if you've got faith in your hand, you've got nothing to worry about." After his surgery he underwent 24 sessions of pulmonary rehab for four hours per day. He had to also undergo chemotherapy to stave off cytomegalovirus (CMV) after surgery.

Now almost 2 years after his surgery, Mr. Steppe's new lungs are functioning at maximum capacity. He is approaching 60 years of age. When I spoke to him this summer, he says he feel great. He walks, plants flowers, and plays with his grandchildren. In one month alone he has ridden 411 miles on his bike. He told me that every time he takes a shower he couldn't help of think of the way he used to be. "You can't help but smile."

David Rice
Liver Recipient, 2002

"I first started getting sick back in 1996. That is when I found out that I had cirrhosis of the liver. Blood wasn't flowing to my liver properly and instead I was bleeding into my stomach. I found myself throwing up blood.

I underwent a procedure to treat my condition, but they told me that I was eventually going to need a new liver. That is when I got on the national waiting list for a new organ.

Two years passed and it looked like I was going to be able to get through it without needing a new liver, but my stomach started bleeding again. My doctors told me that I would need to get a new liver right away or else I didn't have much hope. At that point, my liver was neither pumping nor cleaning my blood.

In addition to my cirrhosis, I had developed a hernia and gallstones. I had to have surgery for this first because my pancreas got infected and I had the aftereffects of a mid-range stroke, which affected my memory.

When you get a liver, they like for you to get it from a relative, if possible. If a relative isn't available, then they start looking at non-related living donors, and then at cadaver donors. That is why my doctors thought my son, David Rice, Jr., who will be 22 this June [2003], might make a good donor.

My son's liver was a perfect match with mine - a six -- but I really didn't want to involve him at all. He flies aircrafts in the Navy and was stationed in Japan. When he found out I needed a liver, however, he said to me, 'I wouldn't be the man I am today if it weren't for you. This is the least I can do for you.'

It got to the point where I was so sick that my wife called the Red Cross. They arranged to have my son flown here from Japan within 24 hours.

The surgery took 13 hours, and they took 51 percent of my son's liver. It was a very successful operation. My doctors were using words to describe it like 'textbook'. They said everything went absolutely smoothly.

I remained in the hospital for about a week. I had to take 14 different medications to keep my body from rejecting my new liver. After two weeks, the swelling went down.

It takes about six weeks for the liver to grow back to its normal size. The operation took place on July 23rd [2002] and my son was able to go back to Japan around Labor Day, after he was cleared through the Bethesda Naval Hospital.

I really appreciated the care and attention that I got at the [University of Maryland] Medical Center. At some places you go, you feel like you are on an assembly line. But everyone at the Medical Center, from my surgeons to the transplant coordinator and all of the nurses, were patient with me and very attentive. They were all so knowledgeable of every detail of my case.

My coordinator would call me in the evenings, on the weekends and on holidays just to check up on me and see how I was doing. For the transplant team at the Medical Center, it isn't just a job. They really care about all of their patients personally."

Marie Joost
Kidney Recipient, 2002

"I have type 1 diabetes. I was diagnosed as a young adult and struggled with it for 23 years before I got my first transplant. I had what was considered brittle diabetes. My blood sugars were always up and down. I was taking four shots of insulin a day and I still couldn't get it under control.

I really started getting sick back in 1994. I was throwing up and there was blood in my urine. I had a lot of symptoms of renal failure. My doctors were getting ready to put me on dialysis when I got a call that they had found me a kidney and a pancreas from a cadaver donor.

After that transplant, they told me that my new organs would probably last me about seven years. Well, in October 2001, I started getting sick again. I was losing a lot of weight and throwing up. My doctor put me on dialysis in December [2001], and I knew I'd need a new organ.

This time around, however, I didn't have to wait for an organ because my husband was willing to give me his kidney. I found out about the University of Maryland's transplant center from my endocrinologist. She referred me to the transplant program.

My recovery has gone pretty smoothly. It took a while for me to regain my strength, but I was able to go back to work part-time in September [2002]. It is good to be back. I feel so much better now than I did before the transplant.

Debbie Schwartz
Pancreas/Kidney Recipient, 2001

"I had diabetes for 31 years, and now I don't. I can't tell you how wonderful that feels.

I'd been monitored for kidney failure for 12 years. They called it creeping failure because it was just creeping along. Things really took a turn for the worse last summer when I wrapped my car around a tree after I blacked out. I was put on an insulin pump after that and my doctors told me to get on the waiting list for a new kidney.

I was only going to have a kidney transplant because that is all that I knew about. I had never heard of the Simultaneous Pancreas Living Kidney (SPLK) procedure. It wasn't until my husband Mark started doing research on kidney transplantation that we learned about it and about the University of Maryland.

We researched transplant facilities nationwide before deciding on Maryland. We were impressed by both the number of procedures that the University of Maryland performed and by their success rate. They have a reputation for getting the surgery done as soon as possible, which is obviously good for the patient.

My experience at the Medical Center was excellent. I was well taken care of. We asked a lot of questions and they answered all of them. Education is so important. You have to be an active participant in your care. You can't just depend on your doctor. Once the transplant team answered our questions thoroughly, we knew that we were in the right place for the right procedure.

My family has a saying that you can't buy good health, but you can buy a better quality of care. I could have stayed in Phoenix and had the transplant done here. I didn't have to fly 2,600 miles across the country to Baltimore. But I might have had to wait another 16 months to have the procedure done, whereas it only took two months for me to get my SPLK at the University of Maryland. And I might not have had both the pancreas and the kidney done at the same time.

As a mother of a five year old and a 12 year old, it was important that I get back to a normal schedule quickly, which I have. My blood sugars have been running under 90 for weeks. I'd had diabetes for as long as I could remember. I lived the first half of my life with it. But now I know that the next half of my life is going to be dramatically different."

Tim Thompson
Liver Donor, 2001

"When I found out that my boss would not survive for more than six months without a liver, I decided to get tested to become a donor. You have to be about the same size as the recipient and have the same blood type. I knew I might be a good match. Once that was established, I had to take a lot of tests.

There were blood tests, an MRI test (magnetic resonance imaging) and psychological tests. They wanted to make sure I had thought through my decision to become a donor.

They showed me pictures of the procedure and gave me lots of information on it. I knew that the procedure would involve them taking a portion of my liver - about half of it - to transplant into my boss. In about six to eight weeks, my liver would grow back to its original size.

For some reason, I was not nervous on the morning of my surgery. I was very calm. The transplant took about 10-1/2 hours. Despite some of the pain involved, I would still do it all over again. There is no greater gratification than saving a life."

Aimee Haugh
Kidney/Pancreas Recipient, 2000

"I found out that I was going to need a new kidney and a new pancreas a couple of years ago, as a result of my diabetes. My doctors suggested that I get both organs transplanted at the same time, instead of having two separate procedures for each organ. That is when my family members and friends started getting tested for a living donor kidney match.

My brother Eric decided to donate his kidney to me, and he was a good match. My brother saved my life. He is my hero. If you have a donor, consider yourself blessed. Every day of my life is a gift.

Twice, the hospital found pancreases for me and had me come in for the transplantation. But right before the surgeries, it turned out that the pancreases weren't really good matches after all. The University of Maryland staff was wonderful throughout everything. They were extraordinary and knowledgeable.

A good pancreas finally came through and the operation was very successful. Before and after the operation were like night and day. It has been unbelievable. I used to wake up with headaches every day. My best days before the transplant can't even compare to my worst days now.

I work full-time, swim two miles a day and have an endless supply of energy. I have so much to be grateful for -- my health, family and friends and the University of Maryland staff for giving me the gift of life again.

The most important thing you can do if you need to have a transplant is to persevere. There is such a difference in the quality of life you can lead after transplantation, and so many opportunities you can take advantage of if you keep a positive attitude."

Eric Schwartz (Aimee's Brother)
Kidney Donor, 2000

"I knew my sister was sick, and knew she would need a kidney. She was on the organ donor waiting list, but I knew that would take some time. I thought the least I could do was get tested. It turned out that I was a good match. Being in such good health, there was no question in my mind that I would donate my kidney to her. It was just a matter of waiting on a good pancreas.

I was glad that I was able to help my sister lead a normal life. The decision to donate is a very serious and important one that people must make for themselves because any time you have surgery, there are risks. But the probability that something would go wrong on my end paled in comparison to the pain my sister was in. My decision to donate my kidney has been extremely fulfilling. It has brought our family closer together than ever.

I was in a fair amount of pain that first week following the procedure, but I was back to work after a couple of weeks. Although I had to increase my activity level slowly, I was able to go jogging and do some light lifting at the gym about six weeks or so after the transplant. I haven't had any problems at all since."

Anthony Winkel
Kidney Recipient, 2000

"My kidneys were getting really bad in the last couple of years before the transplant. I was suffering from end-stage kidney disease. Because of my background in veterinary medicine, I understood what my lab reports meant and was well aware of how much I needed a new kidney.

I didn't know anything about the laparoscopic living donor kidney procedure at that time, but I learned about it from the University of Maryland Web site. I emailed the Transplant Center for more information, and found the staff to be very responsive. I had lots of questions about the procedure, and they answered my questions right away.

My brother Jack was screened and chosen as a match to donate his kidney. We moved forward quickly from there.

Right after the transplantation, the color came back to my face, my appetite picked up, I had energy and an overall increased sense of well-being. When I went home, I felt comfortable. I had good post-operation instructions from my transplant coordinator, and a number to call 24 hours a day in case I had any questions.

I feel so much better now that the surgery is behind me. I don't have to do dialysis anymore, and don't worry as much. I can be really active now without getting tired. I am really happy about it."

Jack Winkel (Anthony's brother)
Kidney Donor, 2000

"Quite frankly, I found that the anxiety leading up to the transplantation was a waste of time. There are days I forget I even donated my kidney to my brother. I lift weights. I play basketball. I am in better shape now than I was before the surgery, and I was only out of work for three days. I wanted to go back to work after two, but my wife wouldn't let me.

My brother was really run down before the transplant. There were a lot of things he wanted to do, but he couldn't. He is so much better now. He is traveling. He is working out. He is like a brand new person. The transplant was good for our whole family."

Roy Bailey
Liver Recipient, 1999

"We have a life again. My wife and I can go on day trips and I can go back to being the handyman around the house that I used to be. I didn't realize how sick I was until I got well. I used to get up in the morning, take medications and sit in a chair all day long. It really wasn't much of a life. I had no energy.

Ironically, I had the liver transplant right on my mother's birthday. I was really sick. And now, I feel really good.

I can't say enough good things about the team at the Medical Center. They were always there to answer all our questions and they were there to support us. It didn't take me long to recuperate and feel like myself again. Everything is better now. And there's so much more we want to do and will be able to do."

Stephen Rash
Kidney Recipient, 1999

"It's so good that I don't have to be tied down to dialysis anymore. The transplant changed my life. It's my third kidney transplant. My last one was in 1987. I feel stronger now and I have more freedom to come and go as I please. I used to have to go to dialysis every Tuesday, Thursday and Saturday. I would be in the middle of something and I'd have to run home in order to get to dialysis. Now, there's a whole life ahead of me. I can travel. I can go away for the weekend.

I tried to make a life for myself even when I was on dialysis, but I used to be so limited. I used to get very sick on Mondays and would feel bad until my dialysis on Tuesday. I couldn't eat or drink certain things. Now, I can go for a coffee and Danish in the morning and not worry about it. People say I look better; I know I feel better. I have a new attitude and outlook on life."

George Raymond
Kidney Recipient, 1998

"I was really suffering the last two years before the transplant. I was close to full kidney failure. I wasn't on dialysis, but I was a day away from it. As I got sicker, my wife decided to donate her kidney to me.

We are both pilots for United Airlines so we weren't sure how much time we would have to take off from work and how that might affect our future. But it only took about three weeks from the time we started our pre-transplant tests to the day of the surgery. They removed my wife's kidney laparoscopically, and she was able to leave the hospital after several days.

The transplant coordinators were outstanding. The doctors were outstanding. Even the hotel where we stayed was nice and clean and right next to the hospital. My wife went back to work first, and then I went back to work after about six months. We've had a good experience. It has been very positive."

Diana Raymond (George's Wife)
Kidney Donor, 1998

"George was put on the waiting list for an organ two years before he finally had his surgery. In that time, we only got one call about a possible kidney that ended up going to someone else on the list.

He really began deteriorating so I insisted on being tested. It turned out we were the same blood type. I knew that was a good starting point. From there, everything went into high gear. I read all of the articles I could find on the laparoscopic procedure. I spoke with a nurse coordinator at the Medical Center, and she was so helpful and enthusiastic. I didn't even know what questions to ask, but she guided me through the process.

The operation only lasted about four and a half hours. They wheeled me in to the recovery room, and I saw George there. The results of the transplant were instantaneous. He had all of his color back. He told me that he felt better than he had in five years."

Shelly D. Arrollo
Kidney recipient, 1999

Hi I'm Shelly the recipient of my best friend Shannon Breshears kidney. This is my story... In august of 1988 I was diagnosed with a rare disease called "polyarteritis nodosa". It's a disease of the arteries and blood vessels in which they expand and burst. The disease attacks the major organs and kills them off. I had blood vessels burst in both of my kidneys and filled one kidney completely with blood and it started filling the second before they were able to stop it. I was in a coma like state for about 2 weeks; I was also given my last rites. I was flown into UCSF because no other hospital would take me because they didn't know enough about the disease. In UCSF I was on a breathing machine because I could not breath on my own. When I finally realized I what was going on I was in shock. It scared me to think I could have died I would have never seen my kids or family again. I was in UCSF for at least a month. UCSF is a teaching hospital I had doctors in my room all the time. The doctors all wanted to know everything about me because of the disease I had, there were at that time only 50 reported cases and all of those patients died within the first 2 weeks. They never made it out of Intensive care (scary to think that could have been me.) I was told on a daily basis that I was going to die; I would probably not live to leave the hospital. If I did get out of the hospital the prognosis was not a good one they gave me 6 months to 1 year to live. I remember laying there (since that's all I could do I couldn't walk) thinking "how would they really know how long I would live they really didn't know anything about this disease. That was 11 years ago and I'm still here and doing GREAT!! My kidney function was stable all this time and I didn't really ever have any serious problems. My creatine was between 2 & 3 forever. I was able to live a normal life, and I didn't have to stop doing anything. It wasn't until 1998 that I started noticed a change. I felt myself getting tired very easily I never really said anything to anyone because I didn't want them to start worrying. My creatine was around 3.5 and higher, I was losing weight and I couldn't keep my food down (I looked like walking death)...

Testing

It was time to start the testing process for the donors. I had 2 who would be testing, Troy, my brother, and Shannon my best friend of 7 years. They received the first kit, which was for the blood tissue typing. They both immediately went to have their blood drawn. We were all on pin and needles. Shannon was never worried she always said from the very beginning she was going to be the one who would give me the kidney. Shannon told me that she believed that that's the reason that we met and became such good friends. She never doubted it was she. I remember she said "wouldn't it be cool if I matched 6 out of 6". I received the phone call from Jeannine my transplant coordinator that Troy was not a match. Troy and I didn't even have the same blood type. I was devastated I couldn't believe it. I remember calling Shannon and telling her that Troy didn't match at all and that it was all up to her (no pressure). Our lives changed from that moment on everything we did was all about the transplant. Jeannine called Shannon and told her that something had happened to her blood and they needed more to complete the tissue. (Shannon said it was because she matched so close they didn't believe it). She wasn't to far off she matched 4 out 6.When I called her to tell her all she could do was just scream she kept saying I told you guys I was the one. I could understand how she could be so happy about losing one of her kidneys. She's a very special person to me and always will be. This is when all the stress started.

Stress

That was the start of the most stressful times I can ever remember in my life. I know it was for Shannon. She felt like everything depended on her. (No pressure there) I was so nervous. Anything could have stopped her from donating. Shannon started all of her testing. She had CXR (chest X-ray), EKG, more blood, urine, and 24-hour urine. She had to do the 24-hour urine twice because her doctor forgot to order a protein test. That is an important test because UCSF needed to know how much protein her kidneys were producing.

While all that was going on, I was getting sicker and sicker. In December '98, I left work on sick leave. My doctor wanted to put me on dialysis. I refused because I wanted to wait to see what happened with Shannon first. In December when they tested Shannon's blood she had Ecoli poisoning. They put her on antibiotics. After that she got strep throat and her doctor wanted to take her tonsils out. She was getting totally stressed. I was ok with it because I knew everything would work out. Finally she all her tests and it was down to the IVP and renal arteriogram. That was a day from HELL.

My parents, Shannon, and I all went to San Francisco so Shannon could do these tests. We were all so stressed. Shannon could not eat. We checked into our room and went to eat. Poor Shannon got to eat Lime Jell-O and to this day she will not eat lime Jell-O. We got up early the next morning to go to the hospital for the final tests. Shannon had her IVP first and I was able to be in the room with her. Following that exam we were off to another floor for her final and most telling exam. Everything hung in the balance as we waited for her renal arteriogram to be completed.

My Recollection

My name is Shannon Breshears, I was born in 1971, and this is the story of my kidney donation experience. On February 17th, 1999, I underwent a donor nephrectomy. The kidney went to my best friend, Shelly Arrollo. Shelly developed a disease around 12 years or so ago that attacked her kidneys. The doctor's told Shelly and her family that she would eventually need a kidney transplant, they just weren't sure when. One day Shelly and I were talking about her health failing and I told her I would test to be a donor when the time came

Testing

I am not going to lie; the testing process was a mental and emotional hell. I have never experienced such a gut-wrenching ordeal. Shelly's brother (Troy) and I decided to test. Everyone thought Troy would match and that I wouldn't because I am not related. We had to wait for approval for the first test (Shelly had HMO insurance). Both Troy and I received packages from UCSF in the mail. We had to go have our blood drawn for a blood/

antigen match. It took what seemed like a month to even hear from the transplant coordinator regarding the results. I remember it was on a Friday evening that Shelly got the call; Troy was not a match. They had different blood types. I had told Shelly from the beginning that I knew I would match. I had never been surer about anything in my life. No one believed it, except for Shelly.

Later in the week, I received another package in the mail from UCSF. I had to have more blood drawn. The excuse was that my blood had gotten contaminated and they needed more. I told Shelly that it was because I matched so well, they had to re-test. I wasn't very far off from the truth. I found out in September or so that I was a 4 out of 6 match for Shelly. This is considered rare, however not impossible. I believe more non-related people would match if they would just test. Hello, it doesn't take a genius to figure that out. I received A LOT of negative feedback when I would tell people what I was doing. It got to the point where I didn't discuss it with anyone except Shelly, her family, and my friend that was a dialysis tech. I was extremely excited and stoked. I had the chance to save my best friends life. She is a sister to me and I couldn't think of a better gift to give. Then there would be these people who would be like, "What if you need your kidney later?" or "What if you have a child and they need your kidney?" Even doctors that we knew were really negative. "There is no way you will match Shannon." "You don't even know if you have two kidney's Shannon." I won't mention any names but they know who they are. All of this negativity really ravaged my spirit at times. It was very hard to keep a positive outlook when I was being suffocated by disdain. I never realized how paralysis of compassion occurs when the fear of the unknown arises. At any rate, this blood drawing turned out to be the easiest portion of the whole testing ordeal.

Ok, so I had just begun the testing journey. After the blood test, I had to go in to see my primary care physician and have a complete physical. I had an EKG and more blood work completed. My memory is fuzzy regarding everything that was done but Shelly kept a journal, so she will be more accurate than I am. I had all sorts of tests. I was tested for HIV and leukemia and I don't know what else. I had to do a 24-hour urine test. My doctor didn't order all of the correct tests so I had to do the 24-hour urine test twice. That was FUN! Yes, I am being facetious of course. I think everyone should get to collect his or her urine in a jug for 24 hours. I had a lot of problems during the testing process as well. I had tonsillitis and strep. Then I got food poisoning over Christmas and UCSF put me on antibiotics for that. I knew I was sick when I did the 24-hour urine the second time, but I thought I had the flu. Janine, the transplant coordinator, called me two weeks later and told me they were putting me on antibiotics. I went along with it even though I wasn't sick at that time.

I am leaving some things out. I had knots in my stomach all of the time. I was so sure that I would match (and I did), but that hospital does not mess around. They will not perform a transplant if they think there will be any danger to the donor. It was so nerve racking. When I finally made it to the last stage of testing, I was an emotional wreck. I had to take an incomplete in one of my classes because the transplant is all I could think about. Shelly was getting sicker and sicker. She was losing weight like crazy and she was throwing up all of the time. I was a mess because she was a mess and I felt like the testing was taking forever. My last phase of testing was the IVP and renal arteriogram. I went to San Francisco with Shelly, her mom (Kathy), and her dad (Louie). We are all snapping at each other and totally grouchy. We were really STRESSED! This was it. Shelly got to go in with me when I had my IVP. This is a test where a nurse injected me with contrast and an x-ray tech took x-rays of my kidneys. It is a little more complex, but not painful. When the films were developed, the tech showed them to us and showed me where my kidneys were on the film. Then we were off to another floor for another test.

The renal arteriogram was a little different. This is where the radiologist inserted a small catheter into my femoral artery (in my upper thigh) and injected dye directly into my renal arteries. This wasn't painful because I was on controlled DRUGS. The experience was very surreal though. I was awake, but on another level of reality. Following the test, I was in recovery and I asked the nurse how it went. The radiologist was addressing concerns before the test and my concern was to know right then if we would be able to do the transplant. So, the doc told me all was well and the nurse reiterated the fact in recovery. I had to stay at UCSF for like 6 hours and keep pressure on the incision. The staff had to monitor me and make sure I didn't bleed from the wound. We left the hospital and felt like we were still in the dark. You have to understand that during this process all of our lives were on hold. We waited for calls from the hospital saying it was ok to do this test or that test, so we felt like we were always playing a waiting game.

I think the call came like a week or so later saying that the transplant was scheduled for the 17th of February. So, we had to wait like two weeks. Talk about trepidation... Ok, we had to check in the day before the transplant. The staff at UCSF told us they would cancel the surgery if anything went wrong. Could they stress me out any more? Any way, we had to check in at 10 or 11 am so we got to have our last meal before we checked in. I rode with Troy (Shelly's brother) and Shelly rode with her boyfriend Joe. It was raining and we stopped to have breakfast on our way to the hospital. I remember the restaurant's grill was broken so they had limited breakfast meals that they could make us. I am getting off track again...

We checked in and the testing and education process was implemented. I won't go into all of the details; you can email me if you really need to know. However, I met my surgeon after he was finished with a liver transplant (very smart guy), and I met my anesthesiologist (hilarious guy). The anesthesiologist looked down my throat, because they were going to stick a tube there during surgery, and told me he was going to give me the equivalent of three martinis. RIGHT ON!!!! Needless to say, I was pretty happy about that. When we first signed all of our paperwork during registration, we had to sign a release. UCSF is a teaching hospital so there are students running around everywhere. It was great though. There was a nursing student that got to watch my part of the transplant and she was amazed. She said that my surgeon handed Shelly's surgeon my kidney and it was like, off he went. The staff kept telling me how much pain I would be in and how much harder it was on the donor than the recipient. This is because of the position of my body on the operating table. I was told that I would be bent in half, on my side. They took the right kidney, which we didn't find out until the day of surgery. They take the kidney that they think will give you problems later in life. My doc told me that often times the right kidney in women will drop later in life and that can cause problems. I felt very secure at that hospital because the staff seemed really knowledgeable and caring. The surgeons thought I only had one artery coming off of my right kidney, but apparently the artery branched off into two. He (Shelly's surgeon) took my kidney and sewed the two arteries into one. Isn't that the coolest? Yeah, it was pretty cool. So, to wrap it up, we were in the hospital for 5 days. I have about a six-inch scar that runs along the rib line and part of my rib missing. I haven't had any problems at all and my labs are normal. Shelly is doing great too, which I am sure she is going to tell you. I don't even have to ever go back to UCSF again. Once a year, I go to my primary care physician and have a few minor tests.

Follow Up from Shannon-August 2003

Hi friends, and newcomers. Here is the latest news about me. Not allot happening actually. I presently reside in Redondo Beach California so I guess I am a SoCal girl, huh. The beach is quite lovely and I am happy to be here. Honestly, the only things I miss from the central valley are my friends. I want to take this opportunity to thank everyone who writes to us. We enjoy hearing from you and communicating with you. If anyone ever wants to write about his or her own experience in organ donation (donor, recipient, family member, friend, etc) let us know. We will dedicate a whole page to you! Take care, Love Shannon and Shelly...aka...Kidney Buddies.

Message from Shelly~April 2003

I have just completed my 4th year of have the best kidney. I still haven't had any problems. I can't even remember when the last time that I was sick with the flu. Shannon was brought into my life for a reason and that was to give me the greatest gift anyone can give another person the gift of life. I still can't believe that it has been 4yrs. I feel great and I really don't have any more side affects to my meds. I'm only on Cellcept, Prograf and Predisone. I still get migraines and they say that is caused by the prograf. Other than that I'm feeling great. It did take a while for me to start feeling 100% but it does happen.

I went for my check-up and the doctor said I was doing great. He told me I could go off of my prednisone but, I would take the risk maybe of going into rejection somewhere down the road. He said it could be 5 or 10 years or maybe never. I didn't think that that was anything thing that I wanted to risk. I'm on a very low dose 5mg every other day. I have no side effect from the prednisone anymore. I believe that your body gets used to it.

Joe and I are still together and it has been almost 5yrs(in Dec.). We are getting ready to move into our new home in a couple of months. Things are going great for me so far. I only have one of my kids at home and he is eighteen. My other son is 21 and lives with his dad. It seems that my life is on track and things are great.

I'm working at Memorial Hospital as a Radiology Clerical Supervisor and I'm doing fine. I sometimes have to work long hours and I have adjusted to that fine. I just take my B12 vitamins and I'm good to go.

Larry McPeak, Liver recipient

Liver recipient Larry McPeak received the "gift of life" two years ago. Now he extends the gift of music and laughter to thousands as the vocalist and bass guitarist for the "VW Boys," a bluegrass trio that combines a unique blend of music, magic and comedy.

The group is on the road regularly throughout the year, from Dollywood in Pigeon Forge, TN, to a recent (July 2) performance at the Kennedy Center in Washington, D.C. The week prior, they represented the music and culture of the Southern Appalachia Region as part of the Smithsonian Folklife Festival on the National Mall.

"Everywhere we perform we stress to our audiences to sign their donor cards and let their family know of their decision," says fellow VW Boy, Tim White. "I, for one, have done that since Larry's transplant. . . nothing, nothing has touched my life like Larry's miracle."

Were they not told his story, his audiences would have no reason to guess that three years ago Larry was holding on to life waiting for a liver transplant. His family and friends, including many musicians, worked through NFT to quickly raise the $35,000 deposit he needed to be placed on the organ donor waiting list in 2000. They then went on to raise a total of approximately $90,000 to pay for transplant-related expenses not covered by insurance.

"It was an outpouring of love; it was just amazing; the entire blue grass community in the U.S. and overseas came together for me," Larry recalls. "Words cannot express the thanks I feel to them and NFT."

Larry has been performing bluegrass music for over 30 years, first with his brothers in the "McPeak Brothers." Later his illness forced him to cut back on his performing. However he joined the VW Boys about six years ago when members Tim White and Dave Vaught "caught me in a strong moment."

Larry took a leave from the VW Boys for 18 months, when he was at his sickest.

But on April 12, 2001, he got a call from the University of Virginia hospital that there was a matching liver. He had the operation the next day, on Good Friday. His doctors were amazed at Larry's fast recovery.

"I had been told that I would be in the hospital three to four weeks, but I was home the next Thursday (less than one week)," Larry says. The following Monday, less than two weeks following his surgery, Larry made it into the studio.

He's been singing with gusto ever since. "I'm having an absolutely fabulous time; it's almost a sin to get paid for doing this," he says. "I am a blessed man."

Larry is a songwriter who has written many bluegrass numbers recorded by other performers. Some of his best-known works include "Seldom Seen," "Country Gentleman," "Lonesome River Band," "Dry Run Creek," "Lady with a Flower in Her Hair," and "Heartaches are in Style this Year." He recently completed a writing project with Tom T. Hall and his wife, Dixie Hall.

Larry's Gospel CD, "There's Always a Calm" includes liner notes that talk about his second chance at life, his faith, and his gratefulness to the doctors, nurses, supporters – and especially to "the young man who lost his life but had the foresight to sign his donor card which enabled me to have a new life."

The VW Boys' newest CD, "Snappy Lunch" is on the Fat Dog Records label. More information about the group can be found at www.vwboys.com.

PEDIATRIC TRANSPLANT SUCCESS STORIES

Kahla Tallie
Year of Birth: 1999

Sheila Williams (Kahla's mother): When Kahla was only 5 months old, she started getting sick. First, she had an ear infection and then a stomach virus. A couple of weeks later, she had to be rushed to the University of Maryland Medical Center because she was having a seizure. That happened only a few days before Christmas.

The doctors found that her blood pressure was very high. They catheterized her to find out why she wasn't passing any urine. They realized her kidney had failed on her. After running a lot of tests, they discovered that something from her stool had made its way to her kidneys and caused them to shut down completely.

Kahla had to stay in the hospital for several months. Her doctors inserted a tube into Kahla's abdomen to give her dialysis. They taught me how to give her dialysis and I was able to take her home, but her blood pressure stayed high and her sonograms showed that her kidneys weren't getting better. In fact, they were getting smaller.

Dr. Mendley talked to me about getting Kahla a kidney transplant. At that time, I didn't know anything about transplants and I couldn't find anything on the Internet about pediatric transplants. The pediatric transplant staff was so helpful. They were always there and they helped me through a lot. Any time I needed someone to talk to, they were there to answer my questions.

I started getting tested to see if I was a match to donate my kidney to Kahla. Different people in my family started to get tested as well, including my mother and Kahla's great aunts. Finally in early December in 2000, I was able to donate my kidney to Kahla. Dr. Farney did the surgery.

Kahla is doing fine now. She is doing really well. She used to go to therapy three times a week, and now she only goes once a week, on Wednesdays. I can't believe it. She is a totally different child now compared to when she was sick.

Madison McClintock
Year of Birth: 1990

Nancy McClintock (Madison's mother): Until Madison became sick; he had no medical problems at all. Then, he started looking pale and he became lethargic. He also started throwing up. On May 10, 2001, I brought him in for a physical and the blood work showed that he needed a kidney and needed to start dialysis.

I was in total shock at that time. I could have taken him to another hospital for his treatment, but I chose the University of Maryland Medical Center because I was more familiar with Maryland. When I was told that Madison was going to need to be on dialysis, I really didn't know anything about it, but I learned.

Madison had to get on peritoneal (through the abdomen) dialysis for nine hours a day while he slept. He was on it for 2 1/2 months until July 27, 2001, when he received a kidney from his father. It was a successful transplant and his kidneys started working immediately. Everything went wonderfully until his creatine level began creeping up. They biopsied his kidney and changed his medication. They tried to help his new kidney to adjust with drugs, but we later found out that he was in acute rejection.

Madison's body had a severe allergic reaction to the antibodies. We are in a holding pattern now to see what his body is doing. He is currently taking 25 pills a day, down from 50 pills. The prednisone in his medications has caused him to gain weight, but he has been a good patient.

Despite his illness, he kept up with all of his work. I ended up home schooling him the last month or so of last school year. He really likes math. In fact, he graduated with honors from his middle school.

Madison is a typical sixth grade boy in many ways. He likes to play basketball and is a big Redskins and World Wrestling Federation fan. Our experience here has been wonderful. Everyone I've met over the last year and a half has been great. They all went out of their way to make sure that he was happy, comfortable and entertained during this process. When he was hospitalized, he was able to play Nintendo and e-mail his friends. I love the staff here. They are so dedicated.

Riley Nelson, 8-year-old Heart Recipient

We all can use new challenges in our lives; and eight year old Riley Nelson of Team Utah at the US Transplant games is no exception.

He has been a participant at the U.S. Transplant Games in the 50-meter dash ('96, '98, '00, '02), bowling ('96, '98, '00, '02), 5K walk ('98), basketball ('02) and Saturday added bicycling to his list of accomplishments.

Riley's eight-year-old transplanted heart was working just fine, thank you, as he pedaled his bike over the 1K time trial course on a Saturday morning, on a beautiful sunny day. Life is good.

Three examples young courage and determination at the US Transplant games before the 1K cycling event.

6 1/2 year-old Gabe Kagan, representing Team Florida, trains on hot dogs, pizza, ice cream, Power Ade, and McDonalds. On the morning of today's race he was up at 5:30 and ate a granola bar.

Gabe prepared specifically for this event by riding his bike "around the block," but he also cross-trained by skateboarding, kung-fu (he has a blue belt), yesterday's 5K walk, swimming at the hotel pool, and playing with his friend, Daniel.

His mom heads Gabe's training team. She is one of a group of coaches accompanying him to the games, and she is the living donor of Gabe's liver on July 11, 1996, when he was 101/2 months old. She's a dedicated coach.

Gabe seems confident but a bit distracted before the race.

Andrew Watts is only 7 years old, but he is meticulous in his preparation. In speaking with this reporter before the race, he was unsure about the handwritten version of his name in my notes, so he insisted on writing it out for me. He also needed to be clear about his age, saying, "I passed 1, I passed 2, I passed 3, I passed 5 -- every single number but 6. Andrew's coach/dad confirmed that Andrew is 7.

Andrew's training regimen includes playing with his SONY, playing with cars, and yes, riding his bike. Though his favorite training foods consist of Italian pasta, ribs and chicken, circumstances on the morning of the race dictated that he postpone his breakfast. "Mom's bringing breakfast," he explained, "and my brother, Justin." They live in Aloma Woods, just outside of Orlando.

Andrew received his heart 2 years ago from a donor in Ohio. This was the culmination of a series of medical procedures that started when he was flown to Boston for the first stage of his surgery when he was only 4 days old. Stages 2 and 3 followed at 6 months and 1 1/2 years. When Andrew was 5 he complained that his feet hurt. Fortunately, his father is a firefighter, trained to check for circulation problems. He went to the hospital and immediately found himself on a medical Lear jet headed for Boston. His new heart, miraculously, was available in a matter of hours. "It was a very close call," his father says. "He was seeing the white light."

Before the race Andrew seemed confident but a bit distracted, circling restlessly on his bike. He was ready.

Andrew's heart was evident during the race. When asked beforehand how far he would ride, he said, "68 miles." His wandering from the course during the race made his prediction truer than anyone thought.

And his infectious smile and energy as he rode his tiny bike aroused the hearts of everyone who saw him.

Six year-old Brooke Miley is technically a "ringer" on Team Alabama, for she and her family live in Mississippi. But since her heart transplant was performed at the University of Alabama-Birmingham where she spent 4 months in the ICU and an additional 2 months recovering, she has a rightful place on her adopted team.

Brooke's coaches were shrugging off their worries about Brooke's recent bout with allergies -- "She coughs like a seal," said her mother and coach. A well conditioned athlete, her favorite foods are apples, grapes, broccoli, "busghetti," "buzzagna," and Coke. Her mom was concerned that her daughter had missed her breakfast, and Brooke seemed a bit distracted by the apple her mother stopped eating to offer her. She asked that Brooke remove her helmet for a picture because, "She has such pretty red hair."

Like many of the younger competitors, Brooke does a lot of cross training. She swims and goes "to the bike trail" near her home. The day before the cycling event she bowled in the games for only the second time in her life, finishing in 5th place.

FREQUENTLY ASKED QUESTIONS

What is transplantation?

Transplantation is the act of surgically removing an organ from one person and placing it into another person. Transplantation occurs because the recipient's organ has failed or has been damaged through illness or injury.

Which organs can be transplanted?

The organs that can be transplanted are:

- liver
- kidney
- pancreas
- kidney/pancreas (can be transplanted at the same time)
- heart
- lung
- heart/lung (can be transplanted at the same time)
- intestine

What policies apply to the transplant I need? How do I learn about potential new policies?

All policies and bylaws governing the Organ Procurement and Transplantation Network can be found under the Resources section of this site and by selecting the appropriate option. Policy 3 contains subsections for all the individual allocation policies.

Organ allocation policies are developed by the OPTN and considered for final approval by the U.S. Department of Health and Human Services (HHS) under federal regulation governing the OPTN.

As new policy proposals are developed by the OPTN for consideration, they will be circulated for public comment. Visit UNOS' Public Comment section to view and respond to such proposals. As part of HHS' consideration of policy proposals advanced by the OPTN, additional public comment may be sought through the Federal Register or other means; this site will also give notice of those actions.

Are there organizations that can help patients afford the cost of transplantation?

Some transplant candidates and recipients have difficulty affording the cost of a transplant or related expenses such as travel and lodging. There are a number of local, regional and national organizations that provide some assistance through grants or services. In individual cases, local community organizations or faith groups may be able to help, and friends and families may solicit funds through public events or appeals. For more information about some of the financial issues transplant candidates and recipients face, as well as available resource information, go to www.transplantliving.org.

What questions should I ask about the cost of transplantation?

- What part of the transplant cost is covered by my insurance?
- What financial coverage is accepted by the hospital?
- How much will I have to pay?
- What happens if my financial coverage runs out?
- Who are the members of the transplant team and what are their jobs?

- How many attending surgeons are available to do my type of transplant?

- Who will tell me about the transplant process?

- Is there a special nursing unit for transplant patients?

- Can I tour the transplant center?

- Will I be asked to take part in research studies?

- Does the hospital do living donor transplants?

- Is a living donor transplant a choice in my case? If so, where will the living donor evaluation be done?

- What is the organ recovery cost if I have a living donor?

Does UNOS handle cord blood or stem cell transplants?

We are involved with the transplantation of vascularized or "solid" organs and are not directly involved with stem cell or cord blood transplants. However, these resources may help you obtain further information on these procedures:

> **National Marrow Donor Program**
> ✉ info@marrow.org
> 🌐 www.marrow.org

> **American Association of Blood Banks**
> ☎ (301) 907-6977
> ✉ aabb@aabb.org
> 🌐 www.aabb.org

> **Fred Hutchinson Cancer Research Center** (FHCRC)
> 🌐 www.fhcrc.org

> **Blood & Marrow Transplant**
> 🌐 www.bmtnews.org

Do transplant hospitals in the U.S. only perform transplants on U.S. citizens?

No. Patients from other countries may travel here to receive transplants. Once accepted by a UNOS transplant center, international patients receive organs based on the same policies as U.S. citizens.

Matching and Allocation

What factors are considered in organ matching and allocation?

Many different medical and logistical characteristics are considered for an organ to be distributed to the best-matched potential recipient. While the specific criteria differ for various organs, matching criteria generally include:

- blood type and size of the organ(s) needed
- time spent awaiting a transplant
- the relative distance between donor and recipient

For certain organs other factors are vital, including:

- the medical urgency of the recipient
- the degree of immune-system match between donor and recipient
- whether the recipient is a child or an adult

For more information, see the Organ Procurement and Transplantation Network article on the Donor Matching System.

How does the matching process work?

The matching process contains six steps:

1. An organ is donated.
2. The donor's information is put into the UNOS transplant information database, UNet^{sm}.
3. UNet^{sm} lists of patients who match that organ.

4. The hospital where the patient is to be transplanted is notified of an available organ.

5. The transplant team considers whether to accept the organ for the patient.

6. The patient who will receive the organ is notified that an organ is available.

To understand how patients are matched on the national waiting list, it's helpful to think of the list as a "pool" of patients. Each time an organ becomes available, UNet^{sm} searches the entire "pool" for the patients who are a match for the organ. A new list is made from those who match.

The patients on this new list are ranked in order of their level of match to that donor organ. The organ is offered to the transplant hospital where the first patient is listed. Other factors, which may be considered, are the patient's current medical status, geographical location, and time on the list. If the organ is refused for any reason, the transplant hospital of the next patient on the list is contacted. This process continues until a match is made.

How do I get on the waiting list?

To get on the national waiting list, you should follow these steps:

o Contact a transplant hospital.There are more than 200 to choose from. You should learn as much as possible about them and choose the one that best meets your needs.

o Make an appointment to visit the hospital. During your visit, the hospital's transplant team will evaluate you (based on your medical history, current condition of health, and other factors) to determine if you would be a good candidate for a transplant. This is called an evaluation.

- o You will have a chance to ask questions. During the evaluation, you should learn as much as possible about that hospital and its transplant team.

- o The hospital's transplant team decides whether you are a good transplant candidate. Each hospital has their own criteria for accepting patients for transplant. You will need to contact each hospital to find out their criteria for accepting patients.

- o If the hospital's transplant team decides that you are a good transplant candidate, they will add you to the UNOS national waiting list.

What are the criteria to be listed?

Each hospital has their own criteria for listing patients. However, UNOS has developed listing guidelines for some organ types.

How do I know that I am listed?

UNOS does not send patients written confirmation of their placement on the waiting list. Instead, patients should find out if they have been placed on the national waiting list through their transplant hospital. If you have questions about your status on the list, you should ask the team at your transplant hospital.

Can I list at more than one hospital?

Yes. UNOS policies permit "multiple listing." However, each hospital has its own criteria for listing patients and may have different rules about patients listing at other hospitals.

How long will I have to wait?

There is no set amount of time, and there is no way to know how long, a patient must wait to receive a donor organ. Factors that affect waiting times are patient medical status, the availability of donors in the local area and the level of match between the donor and recipient.

How will they find the right donor for me?

When a transplant hospital adds you to the waiting list, it is placed in a pool of names. When an organ donor becomes available, all the patients in the pool are compared to that donor. Factors such as medical urgency, time spent on the waiting list, organ size, blood type and genetic makeup are considered. The organ is offered first to the candidate that is the best match.

How are organs distributed?

The organs are distributed locally first, and if no match is found they are then offered regionally, and then nationally, until a recipient is found. Every attempt is made to place donor organs.

Donation

How does whole body donation differ from organ donation? How can I get more information?

Whole body donation for research is handled differently from organ and/or tissue donation for transplantation. Often such arrangements are made through individual medical schools. There are a number of Internet resources that describe whole body donation, including the University of Florida's site on Body Donation Programs in the United States.

What is involved in becoming a living donor? Are there resources that describe the process?

Living donation is arranged through individual transplant centers according to protocols they have set.

Another resource for living donors or those considering living donation is the National Kidney Foundation's National Donor Family Council. Approximately 500 members of the National Donor Family Council and TransAction Council are living donors. Currently, they are working to develop special programs, resources, and features designed to meet the specific needs of living donors and their families. For more information, contact:

National Kidney Foundation's National Donor Council
☎ (800) 622-9010
☎ (212) 889-2210
✉ donorfamily@kidney.org
🌐 www.kidney.org/recips/donor

How does someone get on the waiting list?

The only way for a patient to get on the national waiting list is to visit a transplant hospital. A physician will make an evaluation (based on medical history, current condition of health, and other factors) and decide if the patient meets the criteria to be listed.

Are there age limits or diseases that rule out organ donation?

For any death where organ donation is a possibility and consent is given, there will be a medical assessment of what organs can be recovered. There are no absolute age limits to organ donation. A handful of medical conditions will rule out organ donation, such as HIV-positive status, actively spreading cancer (except for primary brain tumors that have not spread beyond the brain stem), or certain severe, current infections. However, for most other diseases or chronic medical conditions, organ donation remains possible.

Unfortunately, many people never indicate their wish to donate because they believe, falsely, that their age or medical condition would not allow them to donate. If you want to save and enhance lives through donation, the most important action you can take is to share your donation decision; if donation is not medically feasible, that determination will be made at the time of death.

How do I express my wishes to become an organ and tissue donor?

First, indicate your intent to be an organ and tissue donor on your driver's license. Also carry an organ donor card. Most importantly, discuss your decision to donate with your family and loved ones.

Why should minorities be concerned about organ donation?

Some diseases of the kidney, heart, lung, pancreas, and liver are found more frequently in racial and ethnic minority populations than in the general population. For example, African Americans, Asian and Pacific Islanders, and Hispanics are three times more likely to suffer from end-stage renal disease than Caucasians. Native Americans are four times more likely than Caucasians to suffer from diabetes. Some of these diseases are best treated through transplantation; others can only be treated through transplantation.

Successful transplantation often is enhanced by the matching of organs between members of the same ethnic and racial group. For example, any patient is less likely to reject a kidney if an individual who is genetically similar donates it. Generally, people are genetically more similar to people of their own ethnicity or race than to people of other races. Therefore, a shortage of organs donated by minorities can contribute to death and longer waiting periods for transplants for minorities.

How can I help increase organ donation?

There are many ways you can help:

- Become a donor, and talk to your family about your decision to share LIFE.

- Promote donation at work, in your community, at your place of worship, and in your civic organizations

- Make a financial contribution to support UNOS' efforts to raise awareness.

PEDIATRIC FAQ

Will my child start improving once he or she has the transplant?

Once the organ starts working, your child will probably feel much better. However, your child will still need to take the drugs for the rest of his or her life so that the organ is not rejected.

How long will the organ last?

It is impossible to predict how long your child's new organ will last. It depends on many factors including keeping to the drug routine. It is usual for children to need at least two transplants in their lifetime.

Where do we get supplies of the drugs?

You will be able to get your child's drugs from your GP as you did before your child had the transplant.

Will my child still need to stick to his or her special diet?

Your child will not need to stick to the special diet, but weight gain can be a problem after a transplant. This is due partly to the effect of the steroids, and possibly to your child feeling well and wanting to eat all those things that were restricted before.

Gaining too much weight can be bad for your child and can cause extra health problems, so it is best to eat sensibly from the start. Sugar free drinks can help with weight control and will also keep your dentist happy! Avoid snacking between meals on crisps and chocolate and eat more fruit and vegetables. A dietitian can advise you on sensible eating.

Will my child need to measure his or her fluid intake as before?

No, this will not be necessary. However, following a transplant it is important to drink plenty of fluids to prevent dehydration. This is particularly important in hot weather or if your child has a high temperature. If your child is unable to drink fluids or has a severe bout of diarrhea, he or she may need to have an intravenous (into a vein) drip to give fluids. Dehydration can damage the organ, so if you are worried about this please contact your doctor or the Transplant Unit for advice.

When will my child be able to go back to school?

Your child will be able to go back to school about six to eight weeks after the transplant unless your transplant team tells you otherwise. It is important to tell the teachers about your child's new organ, although it should not stop him or her taking part in normal classroom activities.

When can my child start playing games or PE?

Your child will be able to take part in most games after about six to eight weeks, except those that involve direct body contact, like boxing, rugby or karate.

Will my child be at risk from infection?

The anti-rejection drugs suppress or "damp down" your child's immune system, which means that he or she is more likely to pick up infections. Most of them will be common infections, which we all suffer from at some point. However, your child may be affected more severely and the infections could last longer than usual. A common infection in people after transplant is the cold sore virus, which can be painful. However, there are many creams and medicines available, which can catch it early and so, make it less painful. Glandular fever can also be a problem so let the Transplant Unit know if your GP suspects your child has it.

If your child comes into contact with chicken pox or shingles, you should ring the Transplant Unit. Your child will need to have his or her antibodies checked by a blood test. If your child has no antibodies to the chicken pox virus, he or she will need an injection of ZIG (Zoster Immune Globulin) for protection. This applies every time your child comes into contact with chicken pox or shingles.

Can my child have the usual immunizations?

Your child will be able to have most of his or her usual immunizations, but there are a few exceptions:

- MMR (Measles, Mumps, Rubella)
- BCG (Tuberculosis)
- Oral Polio (if in a live form)
- Yellow Fever
- Oral Typhoid (if in a live form)

If you have any questions about immunizations, please ring the Transplant Unit for advice.

Are we entitled to extra benefits?

While you are in hospital, you will be able to meet the social worker. He or she is experienced in helping families who have a child with a transplant and will be able to tell you which benefits you can claim and any other sources of support. It is unlikely that once your child has had a transplant, he or she will be eligible for any extra benefits.

Will we be able to go on vacation?

Your child may need extra vaccinations or medicines when visiting some overseas countries. You should always check in advance which vaccinations or medicines your child needs, and whether he or she is allowed them. The Transplant Unit will be able to advise you about this.

It is always a good idea to take a covering letter with you, outlining your child's medical history in case he or she becomes unwell and needs medical treatment. Always make sure you have adequate medical insurance cover and take enough medicines to cover the period you are away, plus a few days' extra just in case.

Exposure to sunlight can cause skin cancer. The risk of this increases in transplant patients because of the drugs they are taking. So it is vital that you protect your child from direct sunlight as far as possible, always use a high protection sun cream of SPF 25 or more, stay in the shade and wear a hat.

Feelings and emotions before and after a transplant

Clearly, your family life is affected by having a child with chronic renal failure who will, at some stage, require dialysis. There are also stresses involved when your child goes onto the waiting list for, and receives, a new kidney.

It is all too easy for families to feel that their family life is dominated by the chronic illness and its treatment, and there will be times when this is unavoidable. However, it is important for all members of the family to have as normal a family life as possible. The illness is important but most of the time it is not the most important part of your lives.

Inevitably, a child with a chronic illness will have his or her parent's attention focused on him or her regularly but it is important that this attention is not always connected to the illness. It is also important for brothers and sisters to feel that they too have some exclusive time with their parents.

Transplant can sometimes feel like the ultimate goal of your child's treatment and once that goal is achieved everyone can relax. This, unfortunately, is not the case. A transplant brings with it an increase in energy and good health as well as freedom from dialysis. However, it is not the end of treatment. Medication has to be taken every day to prevent rejection of the kidney and an adequate fluid intake has to be guaranteed. This will ensure that the transplanted kidney has as long a life as possible.

Again, although it is important to adhere to treatment it is important that the transplant assumes its proper place in family life. At times it will be at the top of your list of priorities and, at other times, it will be some way down that list.

Although you, as parents, will understand the importance of taking medication and drinking adequate fluids, these will, at times, seem like a burden to your child. This is understandable, and the 'burden' has to be shared if it is not to become overwhelming. The sharing can be done by you as parents but there are many members of the hospital team who will be able to offer different kinds of help and expertise. Please feel free to seek help at any time with any problem or difficulty you may be experiencing.

Non-adherence to a treatment regimen can be a particular problem during adolescence. Naturally enough, teenagers only want to be 'normal'. Being different can be intolerable. This may seem especially difficult immediately after transplant when the effects of medication are obvious. These may include an increase in weight and/or bodily and facial hair. These effects do diminish in time as medication is reduced and the new kidney is in less danger of being rejected. It is important at this time to continue to support your son or daughter despite his or her demands to be completely independent.

The move towards adulthood can be difficult for both the teenager and his or her parents. Although it may be tempting to let your bad-tempered teenager look after him or herself, this is not a good idea. Alternatively, doing everything for your rather irritable 14 year old to prevent mishap is also not recommended! A balance has to be struck and a partnership approach adopted, with your son or daughter gradually taking more and more responsibility so that at the age of 16 he or she can act independently. Medical and nursing staff is always there to help your family through what can sometimes be a difficult process.

It is important that children are properly prepared for the transition from a children's to an adult hospital. This transition can also be a difficult time for you as parents. The move to independence on your child's part may be welcome but may also rob you of a certain intimacy with your child. You may even find yourself missing the chore of dialysis or the role of medication-giver and fluid monitor! These are normal feelings after looking after a dependent child for so long and part of this transition is finding other things to share besides kidney disease and its treatment.

COMMON MYTHS OF ORGAN DONATION

————— ✢ —————

There is a severe organ shortage in this country. Despite continuing efforts at public education, misconceptions and inaccuracies about donation persist. It's a tragedy if even one person decides against donation because they don't know the truth. Following is a list of the most common myths along with the actual facts:

Myth: If emergency room doctors know you're an organ donor, they won't work as hard to save you.

Fact: If you are sick or injured and admitted to the hospital, the number one priority is to save your life. Organ donation can only be considered after a physician has declared brain death. Many states have adopted legislation allowing individuals to legally designate their wish to be a donor should brain death occur, although in many states Organ Procurement Organizations also require consent from the donor's family.

Myth: When you're waiting for a transplant, your financial or celebrity status is as important as your medical status.

Fact: When you are on the transplant waiting list for a donor organ, what really counts is the severity of your illness, time spent waiting, blood type, and other important medical information.

Myth: Having "organ donor" noted on your driver's license or carrying a donor card is all you have to do to become a donor.

Fact: While a signed donor card and a driver's license with an "organ donor" designation are legal documents, organ and tissue donation is usually discussed with family members prior to the donation. To ensure that your family understands your wishes, it is important that you tell your family about your decision to donate LIFE.

Myth: Only hearts, livers, and kidneys can be transplanted.

Fact: Needed organs include the heart, kidneys, pancreas, lungs, liver and intestines. Tissues that can be donated include the eyes, skin, bone, heart valves and tendons.

Myth: Your history of medical illness means your organs or tissues are unfit for donation.

Fact: At the time of death, the appropriate medical professionals will review your medical and social histories to determine whether or not you can be a donor. With recent advances in transplantation, many more people than ever before can be donors. It's best to tell your family your wishes and sign up to be an organ and tissue donor on your driver's license or an official donor document.

Myth: You are too old to be a donor.

Fact: People of all ages and medical histories should consider themselves potential donors. Your medical condition at the time of death will determine what organs and tissue can be donated.

Myth: If you agree to donate your organs, your family will be charged for the costs.

Fact: There is no cost to the donor's family or estate for organ and tissue donation. Funeral costs remain the responsibility of the family.

Myth: Organ donation disfigures the body and changes the way it looks in a casket.

Fact: Donated organs are removed surgically, in a routine operation similar to gallbladder or appendix removal. Donation does not change the appearance of the body for the funeral service.

Myth: Your religion prohibits organ donation.

Fact: All major organized religions approve of organ and tissue donation and consider it an act of charity.

Myth: There is real danger of being heavily drugged, then waking to find you have had one kidney (or both) removed for a black market transplant.

Fact: This tale has been widely circulated over the Internet. There is absolutely no evidence of such activity ever occurring in the U.S. While the tale may sound credible, it has no basis in the reality of organ transplantation. Many people who hear the myth probably dismiss it, but it is possible that some believe it and decide against organ donation out of needless fear.

BIBLIOGRAPHY

www.kidney.org.uk/Medical-Info/glossary/glossary.html

www.niddk.nih.gov/health/diabetes/pubs/dmdict/K-O.htm

http://www.kidneywise.com/basics/kidneys/causes.asp

http://www.diabetes.org/type-1-diabetes/pancreas-transplants.jsp

Port, Friedrich K. et. al. "Comparison of Survival Probabilities for Dialysis Patients vs. Cadaveric Renal Transplant Recipients." *Journal of the American Medical Association.* Vol. 270 (1993), pp. 1339-1343.

Stratta, Robert J. et. al. "The Analysis of Benefit and Risk of Combined Pancreatic and Renal Transplantation versus Renal Transplantation Alone." *Surgery, Gynecology, and Obstetrics.* Vol. 177 (1993), pp. 163-171.

http://www.clevelandclinic.org/health/health-info/docs/0200/0236.asp?index=4754&src=news

http://www.transplantliving.org/afterthetransplant/dietAndExercise.aspx

http://www.transplantliving.org/afterthetransplant/

http://www.ustransplant.org/cgi-bin/urrea?p=index.php

http://www.ich.ucl.ac.uk/factsheets/test_procedure_operations/kidney_transplant/index.html

http://www.unos.org/

http://medlineplus.gov/

http://www.transweb.org/webcast/usa2004/index.htm

Dealing With your Diagnosis: Portions reprinted with permission. © 2004 WarrenShepell

Stories used with permission by University of Maryland Medical System Web Site.
http://www.umm.edu/transplant/testimonials.html

Graphics: Gray's Anatomy of the Human Body
http://www.bartleby.com/107/

ABOUT THE AUTHOR

——————— ⚜ ———————

Jeff is a board-certified family nurse practitioner. His research experience includes numerous cardiovascular and diabetes trials. He has been published in numerous journals, including the LANCET and The American Journal of Cardiology.

Jeff has served on the Executive Committee for the University of Tennessee, Memphis. He is a member of the American Nurses Association, Tennessee Nurses Association, The American College of Nurse Practitioners, Sigma Theta Tau Honor Society, United Who's Who for Professionals and Executives Worldwide, and the Imhotep Society. He is on the Board of Directors, Chester County Senior Center. He also serves on a Clinical Advisory Board for PhRMA in Washington, D.C.

Jeff received his undergraduate degrees in Nursing from Union University. His graduate degree was completed at the University of Tennessee, Memphis. His hobbies are reading, playing drums, and playing golf. He also has a passion for politics and has worked on numerous local and national campaigns.

He is currently in private practice in Henderson, TN. His clinic provides: diagnosis and treatment of major and minor illness; annual check-ups physical exams – school, sports, employment; women's services including pap smears, breast exams, PMS, menopausal problems; minor surgery and minor emergencies; cancer screening and onsite diagnostic testing, EKG, vision, hearing, and laboratory services; provides pediatric through geriatric care: provides birth control information and services; and also provides preventative care and health maintenance services.

www.ingramcontent.com/pod-product-compliance
Lightning Source LLC
Chambersburg PA
CBHW020441290526
45785CB00002B/965